GCSE AQA
Physics
The Revision Guide

This book is for anyone doing **GCSE AQA Physics**.
It covers everything you'll need for your year 10 and 11 exams.

GCSE Science is all about **understanding how science works**.
And not only that — understanding it well enough to be able to **question** what you hear on TV and read in the papers.

But you can't do that without a fair chunk of **background knowledge**. Hmm, tricky.

Happily this CGP book includes all the **science facts** you need to learn, and shows you how they work in the **real world**. And in true CGP style, we've explained it all as **clearly and concisely** as possible.

It's also got some daft bits in to try and make the whole experience at least vaguely entertaining for you.

What CGP is all about

Our sole aim here at CGP is to produce the highest quality books — carefully written, immaculately presented and dangerously close to being funny.

Then we work our socks off to get them out to you — at the cheapest possible prices.

Contents

PHYSICS 2B — ELECTRICITY AND THE ATOM

PHYSICS 3A — MEDICAL APPLICATIONS OF PHYSICS

PHYSICS 3B — FORCES AND ELECTROMAGNETISM

Published by CGP

From original material by Richard Parsons.

Editors:
Ellen Bowness, Helena Hayes, Felicity Inkpen, Edmund Robinson,
Hayley Thompson, Julie Wakeling, Sarah Williams.

Contributors:
Paddy Gannon, Sandy Gardner, Gemma Hallam.

ISBN: 978 1 84762 627 1

With thanks to Mark A Edwards, Martin Payne, Glenn Rogers, Karen Wells
and Dawn Wright for the proofreading.
With thanks to Jan Greenway for the copyright research.

With thanks to iStockphoto.com for use of the images on page 83.

Groovy website: www.cgpbooks.co.uk

Printed by Elanders Ltd, Newcastle upon Tyne.
Jolly bits of clipart from CorelDRAW®

The Scientific Process

You need to know a few things about how the world of science works. First up is the scientific process — how a scientist's mad idea turns into a widely accepted theory.

Scientists Come Up with Hypotheses — Then Test Them

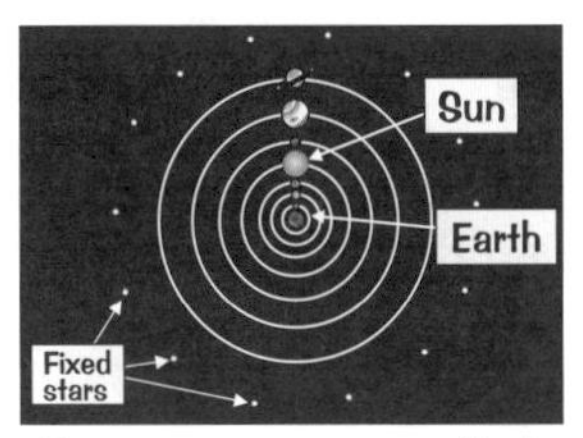

About 500 years ago, we still thought the Solar System looked like this.

1) Scientists try to explain things. Everything.
2) They start by observing something they don't understand — it could be anything, e.g. planets in the sky, a person suffering from an illness, what matter is made of... anything.
3) Then, they come up with a hypothesis — a possible explanation for what they've observed.
4) The next step is to test whether the hypothesis might be right or not — this involves gathering evidence (i.e. data from investigations).
5) The scientist uses the hypothesis to make a prediction — a statement based on the hypothesis that can be tested. They then carry out an investigation.
6) If data from experiments or studies backs up the prediction, you're one step closer to figuring out if the hypothesis is true.

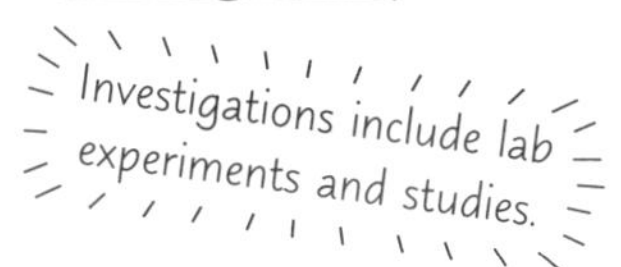

Other Scientists Will Test the Hypothesis Too

1) Other scientists will use the hypothesis to make their own predictions, and carry out their own experiments or studies.
2) They'll also try to reproduce the original investigations to check the results.
3) And if all the experiments in the world back up the hypothesis, then scientists start to think it's true.
4) However, if a scientist somewhere in the world does an experiment that doesn't fit with the hypothesis (and other scientists can reproduce these results), then the hypothesis is in trouble.
5) When this happens, scientists have to come up with a new hypothesis (maybe a modification of the old hypothesis, or maybe a completely new one).

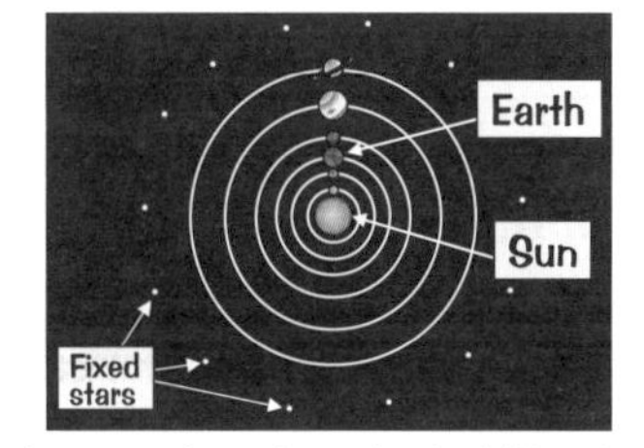

Then we thought it looked like this.

If Evidence Supports a Hypothesis, It's Accepted — for Now

1) If pretty much every scientist in the world believes a hypothesis to be true because experiments back it up, then it usually goes in the textbooks for students to learn.
2) Accepted hypotheses are often referred to as theories.

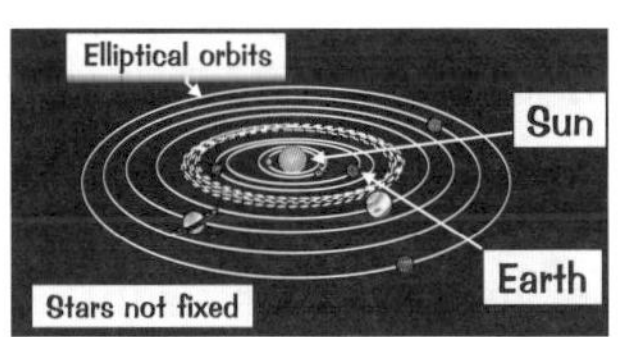

Now we think it's more like this.

3) Our currently accepted theories are the ones that have survived this 'trial by evidence' — they've been tested many, many times over the years and survived (while the less good ones have been ditched).
4) However... they never, never become hard and fast, totally indisputable fact. You can never know... it'd only take one odd, totally inexplicable result, and the hypothesising and testing would start all over again.

You expect me to believe that — then show me the evidence...

If scientists think something is true, they need to produce evidence to convince others — it's all part of testing a hypothesis. One hypothesis might survive these tests, while others won't — it's how things progress. And along the way some hypotheses will be disproved — i.e. shown not to be true.

Your Data's Got To be Good

Evidence is the key to science — but not all evidence is equally good.
The way evidence is gathered can have a big effect on how trustworthy it is...

Lab Experiments and Studies Are Better Than Rumour

1) Results from experiments in laboratories are great. A lab is the easiest place to control variables so that they're all kept constant (except for the one you're investigating). This makes it easier to carry out a FAIR TEST.
2) For things that you can't investigate in the lab (e.g. climate) you conduct scientific studies. As many of the variables as possible are controlled, to make it a fair test.
3) Old wives' tales, rumours, hearsay, "what someone said", and so on, should be taken with a pinch of salt. Without any evidence they're NOT scientific — they're just opinions.

The Bigger the Sample Size the Better

1) Data based on small samples isn't as good as data based on large samples. A sample should be representative of the whole population (i.e. it should share as many of the various characteristics in the population as possible) — a small sample can't do that as well.
2) The bigger the sample size the better, but scientists have to be realistic when choosing how big. For example, if you were studying how lifestyle affects people's weight it'd be great to study everyone in the UK (a huge sample), but it'd take ages and cost a bomb. Studying a thousand people is more realistic.

Evidence Needs to be Reliable (Repeatable and Reproducible)

Evidence is only reliable if it can be repeated (during an experiment) AND other scientists can reproduce it too (in other experiments). If it's not reliable, you can't believe it.

RELIABLE means that the data can be repeated, and reproduced by others.

EXAMPLE: In 1989, two scientists claimed that they'd produced 'cold fusion' (the energy source of the Sun — but without the big temperatures). It was huge news — if true, it would have meant cheap and abundant energy for the world... forever. However, other scientists just couldn't reproduce the results — so the results weren't reliable. And until they are, 'cold fusion' isn't going to be accepted as fact.

Evidence Also Needs to Be Valid

VALID means that the data is reliable AND answers the original question.

EXAMPLE: DO POWER LINES CAUSE CANCER?
Some studies have found that children who live near overhead power lines are more likely to develop cancer. What they'd actually found was a correlation (relationship) between the variables "presence of power lines" and "incidence of cancer" — they found that as one changed, so did the other.
But this evidence is not enough to say that the power lines cause cancer, as other explanations might be possible. For example, power lines are often near busy roads, so the areas tested could contain different levels of pollution from traffic. So these studies don't show a definite link and so don't answer the original question.

RRRR — Remember, Reliable means Repeatable and Reproducible...

By now you should have realised how important trustworthy evidence is (even more important than a good supply of spot cream). Unfortunately, you need to know loads more about fair tests and experiments — see p. 5-10.

Bias and Issues Created by Science

It isn't all hunky-dory in the world of science — there are some problems...

Scientific Evidence can be Presented in a Biased Way

1) People who want to make a point can sometimes present data in a biased way, e.g. they overemphasise a relationship in the data. (Sometimes without knowing they're doing it.)
2) And there are all sorts of reasons why people might want to do this — for example...

- They want to keep the organisation or company that's funding the research happy. (If the results aren't what they'd like they might not give them any more money to fund further research.)
- Governments might want to persuade voters, other governments, journalists, etc.
- Companies might want to 'big up' their products. Or make impressive safety claims.
- Environmental campaigners might want to persuade people to behave differently.

Things can Affect How Seriously Evidence is Taken

1) If an investigation is done by a team of highly-regarded scientists it's sometimes taken more seriously than evidence from less well known scientists.
2) But having experience, authority or a fancy qualification doesn't necessarily mean the evidence is good — the only way to tell is to look at the evidence scientifically (e.g. is it reliable, valid, etc.).
3) Also, some evidence might be ignored if it could create political problems, or emphasised if it helps a particular cause.

EXAMPLE: Some governments were pretty slow to accept the fact that human activities are causing global warming, despite all the evidence. This is because accepting it means they've got to do something about it, which costs money and could hurt their economy. This could lose them a lot of votes.

Scientific Developments are Great, but they can Raise Issues

Scientific knowledge is increased by doing experiments. And this knowledge leads to scientific developments, e.g. new technologies or new advice. These developments can create issues though. For example:

Economic issues: Society can't always afford to do things scientists recommend (e.g. investing heavily in alternative energy sources) without cutting back elsewhere.

Social issues: Decisions based on scientific evidence affect people — e.g. should fossil fuels be taxed more highly (to invest in alternative energy)? Should alcohol be banned (to prevent health problems)? Would the effect on people's lifestyles be acceptable...

Environmental issues: Nuclear power stations can provide us with a reliable source of electricity, but disposing of the waste can lead to environmental issues.

Ethical issues: There are a lot of things that scientific developments have made possible, but should we do them? E.g. develop better nuclear weapons.

Trust me — I've got a BSc, PhD, PC, TV and a DVD...

We all tend to swoon at people in authority, but you have to ignore that fact and look at the evidence (just because someone has got a whacking great list of letters after their name doesn't mean the evidence is good). Spotting biased evidence isn't the easiest thing in the world — ask yourself 'Does the scientist (or the person writing about it) stand to gain something (or lose something)?' If they do, it's possible that it could be biased.

Science Has Limits

Science can give us amazing things — cures for diseases, space travel, heated toilet seats...
But science has its limitations — there are questions that it just can't answer.

Some Questions Are Unanswered by Science — So Far

1) We don't understand everything. And we never will. We'll find out more, for sure — as more hypotheses are suggested, and more experiments are done. But there'll always be stuff we don't know.

EXAMPLES:

- Today we don't know as much as we'd like about the impacts of global warming. How much will sea level rise? And to what extent will weather patterns change?
- We also don't know anywhere near as much as we'd like about the Universe. Are there other life forms out there? And what is the Universe made of?

2) These are complicated questions. At the moment scientists don't all agree on the answers because there isn't enough reliable and valid evidence.
3) But eventually, we probably will be able to answer these questions once and for all... All we need is more evidence.
4) But by then there'll be loads of new questions to answer.

Other Questions Are Unanswerable by Science

1) Then there's the other type... questions that all the experiments in the world won't help us answer — the "Should we be doing this at all?" type questions. There are always two sides...
2) Take space exploration. It's possible to do it — but does that mean we should?
3) Different people have different opinions.

For example...
Some people say it's a good idea... it increases our knowledge about the Universe, we develop new technologies that can be useful on Earth too, it inspires young people to take an interest in science, etc.
Other people say it's a bad idea... the vast sums of money it costs should be spent on more urgent problems, like providing clean drinking water and curing diseases in poor countries. Others say that we should concentrate research efforts on understanding our own planet better first.

4) The question of whether something is morally or ethically right or wrong can't be answered by more experiments — there is no "right" or "wrong" answer.
5) The best we can do is get a consensus from society — a judgement that most people are more or less happy to live by. Science can provide more information to help people make this judgement, and the judgement might change over time. But in the end it's up to people and their conscience.

Chips or rice? — totally unanswerable by science...

Right — get this straight in your head — science can't tell you whether you should or shouldn't do something. That kind of thing is up to you and society to decide. There are tons of questions that science might be able to answer in the future — like how much sea level might rise due to global warming, what the Universe is made of and whatever happened to those pink stripy socks with Santa on that I used to have.

Designing Investigations

Dig out your lab coat and dust down your badly-scratched safety goggles... it's investigation time. You need to know a shed load about investigations for your controlled assessment and all your exams. Investigations include experiments and studies. The next six pages take you from start to finish. Enjoy.

Investigations Produce Evidence to Support or Disprove a Hypothesis

1) Scientists observe things and come up with hypotheses to explain them (see page 1).
2) To figure out whether a hypothesis might be correct or not you need to do an investigation to gather some evidence.
3) The first step is to use the hypothesis to come up with a prediction — a statement about what you think will happen that you can test.
4) For example, if your hypothesis is:

   "Spots are caused by picking your nose too much."

   Then your prediction might be:

   "People who pick their nose more often will have more spots."

5) Investigations are used to see if there are patterns or relationships between two variables. For example, to see if there's a pattern or relationship between the variables 'having spots' and 'nose picking'.
6) The investigation has to be a FAIR TEST to make sure the evidence is reliable and valid...

Sometimes the words 'hypothesis' and 'prediction' are used interchangeably.

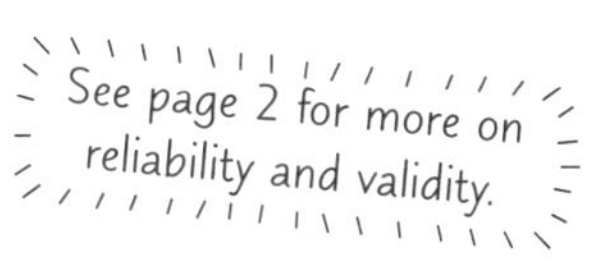

To Make an Investigation a Fair Test You Have to Control the Variables

1) In a lab experiment you usually change one variable and measure how it affects the other variable.

   EXAMPLE: you might change only the angle of a slope and measure how it affects the time taken for a toy car to travel down it.

2) To make it a fair test everything else that could affect the results should stay the same (otherwise you can't tell if the thing you're changing is causing the results or not — the data won't be reliable or valid).

   EXAMPLE continued: you need to keep the slope length the same, otherwise you won't know if any change in the time taken is caused by the change in angle, or the change in length.

3) The variable you CHANGE is called the INDEPENDENT variable.
4) The variable you MEASURE is called the DEPENDENT variable.
5) The variables that you KEEP THE SAME are called CONTROL variables.

EXAMPLE continued:
Independent variable = angle of slope
Dependent variable = time taken
Control variable = length of slope

Designing Investigations

Trial Runs help Figure out the Range and Interval of Variable Values

1) It's a good idea to do a trial run first — a quick version of your experiment.
2) Trial runs are used to figure out the range of variable values used in the proper experiment (the upper and lower limit). For example, if you can't accurately measure the change in the dependent variable at the upper values in the trial run, you might narrow the range in the proper experiment.

Slope example from previous page continued:

- You might do trial runs at 20, 40, 60 and 80°. If the time taken is too short to accurately measure at 80°, you might narrow the range to 20-60°.
- If using 20° intervals gives you a big change in time taken you might decide to use 10° intervals, e.g. 20, 30, 40, 50°...

3) And trial runs can be used to figure out the interval (gaps) between the values too. The intervals can't be too small (otherwise the experiment would take ages), or too big (otherwise you might miss something).
4) Trial runs can also help you figure out how many times the experiment has to be repeated to get reliable results. E.g. if you repeat it three times and the results are all similar, then three repeats is enough.

It Can Be Hard to Control the Variables in a Study

It's important that a study is a fair test, just like a lab experiment. It's a lot trickier to control the variables in a study than it is in a lab experiment though (see previous page). Sometimes you can't control them all, but you can use a control group to help. This is a group of whatever you're studying (people, plants, lemmings, etc.) that's kept under the same conditions as the group in the experiment, but doesn't have anything done to it.

EXAMPLE: If you're studying the effect of pesticides on crop growth, pesticide is applied to one field but not to another field (the control field). Both fields are planted with the same crop, and are in the same area (so they get the same weather conditions). The control field is there to try and account for variables like the weather, which don't stay the same all the time, but could affect the results.

Investigations Can be Hazardous

1) A hazard is something that can potentially cause harm. Hazards include:

- Microorganisms, e.g. some bacteria can make you ill.
- Chemicals, e.g. sulfuric acid can burn your skin and alcohols catch fire easily.
- Fire, e.g. an unattended Bunsen burner is a fire hazard.
- Electricity, e.g. faulty electrical equipment could give you a shock.

2) Scientists need to manage the risk of hazards by doing things to reduce them. For example:

- If you're working with sulfuric acid, always wear gloves and safety goggles. This will reduce the risk of the acid coming into contact with your skin and eyes.
- If you're using a Bunsen burner, stand it on a heat proof mat. This will reduce the risk of starting a fire.

You can find out about potential hazards by looking in textbooks, doing some internet research, or asking your teacher.

You won't get a trial run at the exam, so get learnin'...

All this info needs to be firmly lodged in your memory. Learn the names of the different variables — if you remember that the variable you chaNge is called the iNdependent variable, you can figure out the other ones.

Collecting Data

After designing an investigation that's so beautiful people will marvel at it for years to come, you'll need to get your hands mucky and collect some data.

Your Data Should be as Reliable, Accurate and Precise as Possible

1) To improve reliability you need to repeat the readings and calculate the mean (average). You need to repeat each reading at least three times.
2) To make sure your results are reliable you can cross check them by taking a second set of readings with another instrument (or a different observer).
3) Checking your results match with secondary sources, e.g. other studies, also increases the reliability of your data.
4) Your data also needs to be ACCURATE. Really accurate results are those that are really close to the true answer.
5) Your data also needs to be PRECISE. Precise results are ones where the data is all really close to the mean (i.e. not spread out).

Repeat	Data set 1	Data set 2
1	12	11
2	14	17
3	13	14
Mean	13	14

Data set 1 is more precise than data set 2.

Your Equipment has to be Right for the Job

1) The measuring equipment you use has to be sensitive enough to measure the changes you're looking for. For example, if you need to measure changes of 1 ml you need to use a measuring cylinder that can measure in 1 ml steps — it'd be no good trying with one that only measures 10 ml steps.
2) The smallest change a measuring instrument can detect is called its RESOLUTION. E.g. some mass balances have a resolution of 1 g, some have a resolution of 0.1 g, and some are even more sensitive.
3) Also, equipment needs to be calibrated so that your data is more accurate. E.g. mass balances need to be set to zero before you start weighing things.

You Need to Look out for Errors and Anomalous Results

1) The results of your experiment will always vary a bit because of random errors — tiny differences caused by things like human errors in measuring.
2) You can reduce their effect by taking many readings and calculating the mean.
3) If the same error is made every time, it's called a SYSTEMATIC ERROR. For example, if you measured from the very end of your ruler instead of from the 0 cm mark every time, all your measurements would be a bit small.
4) Just to make things more complicated, if a systematic error is caused by using equipment that isn't calibrated properly it's called a ZERO ERROR. For example, if a mass balance always reads 1 gram before you put anything on it, all your measurements will be 1 gram too heavy.
5) You can compensate for some systematic errors if you know about them though, e.g. if your mass balance always reads 1 gram before you put anything on it you can subtract 1 gram from all your results.
6) Sometimes you get a result that doesn't seem to fit in with the rest at all.
7) These results are called ANOMALOUS RESULTS.
8) You should investigate them and try to work out what happened. If you can work out what happened (e.g. you measured something totally wrong) you can ignore them when processing your results.

Repeating the experiment in the exact same way and calculating an average won't correct a systematic error.

Park	Number of pigeons	Number of crazy tramps
A	28	1
B	42	2
C	1127	0

Zero error — sounds like a Bruce Willis film...

Weirdly, data can be really precise but not very accurate, e.g. a fancy piece of lab equipment might give results that are precise, but if it's not calibrated properly those results won't be accurate.

Processing and Presenting Data

After you've collected your data you'll have oodles of info that you have to make some kind of sense of. You need to process and present it so you can look for patterns and relationships in it.

Data Needs to be Organised

1) Tables are dead useful for organising data.
2) When you draw a table use a ruler, make sure each column has a heading (including the units) and keep it neat and tidy.
3) Annoyingly, tables are about as useful as a chocolate teapot for showing patterns or relationships in data. You need to use some kind of graph for that.

You Might Have to Process Your Data

1) When you've done repeats of an experiment you should always calculate the mean (average). To do this ADD TOGETHER all the data values and DIVIDE by the total number of values in the sample.
2) You might also need to calculate the range (how spread out the data is). To do this find the LARGEST number and SUBTRACT the SMALLEST number from it.

Ignore anomalous results when calculating these.

EXAMPLE

Test tube	Repeat 1 (g)	Repeat 2 (g)	Repeat 3 (g)	Mean (g)	Range (g)
A	28	37	32	(28 + 37 + 32) ÷ 3 = 32.3	37 – 28 = 9
B	47	51	60	(47 + 51 + 60) ÷ 3 = 52.7	60 – 47 = 13
C	68	72	70	(68 + 72 + 70) ÷ 3 = 70.0	72 – 68 = 4

If Your Data Comes in Categories, Present It in a Bar Chart

1) If the independent variable is categoric (comes in distinct categories, e.g. blood types, metals) you should use a bar chart to display the data.
2) You also use them if the independent variable is discrete (the data can be counted in chunks, where there's no in-between value, e.g. number of people is discrete because you can't have half a person).
3) There are some golden rules you need to follow for drawing bar charts:

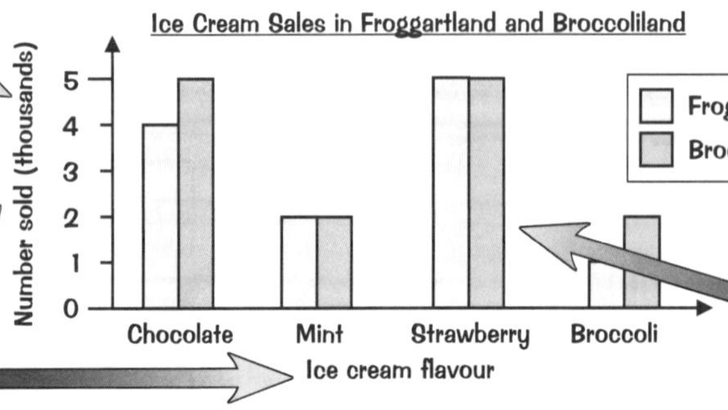

Remember to include the units.

Label both axes.

If you've got more than one set of data include a key.

Draw it nice and big (covering at least a third of the graph paper).

Leave a gap between different categories.

Discrete variables love bar charts — although they'd never tell anyone that...

The stuff on this page might all seem a bit basic, but it's easy marks in the exams (which you'll kick yourself if you don't get). Examiners are a bit picky when it comes to bar charts — if you don't draw them properly they won't be happy. Also, double check any mean or range calculations you do, just to be sure they're correct.

Presenting Data

Scientists just love presenting data as line graphs (weirdos)...

If Your Data is Continuous, Plot a Line Graph

1) If the independent variable is continuous (numerical data that can have any value within a range, e.g. length, volume, temperature) you should use a line graph to display the data.
2) Here are the rules for drawing line graphs:

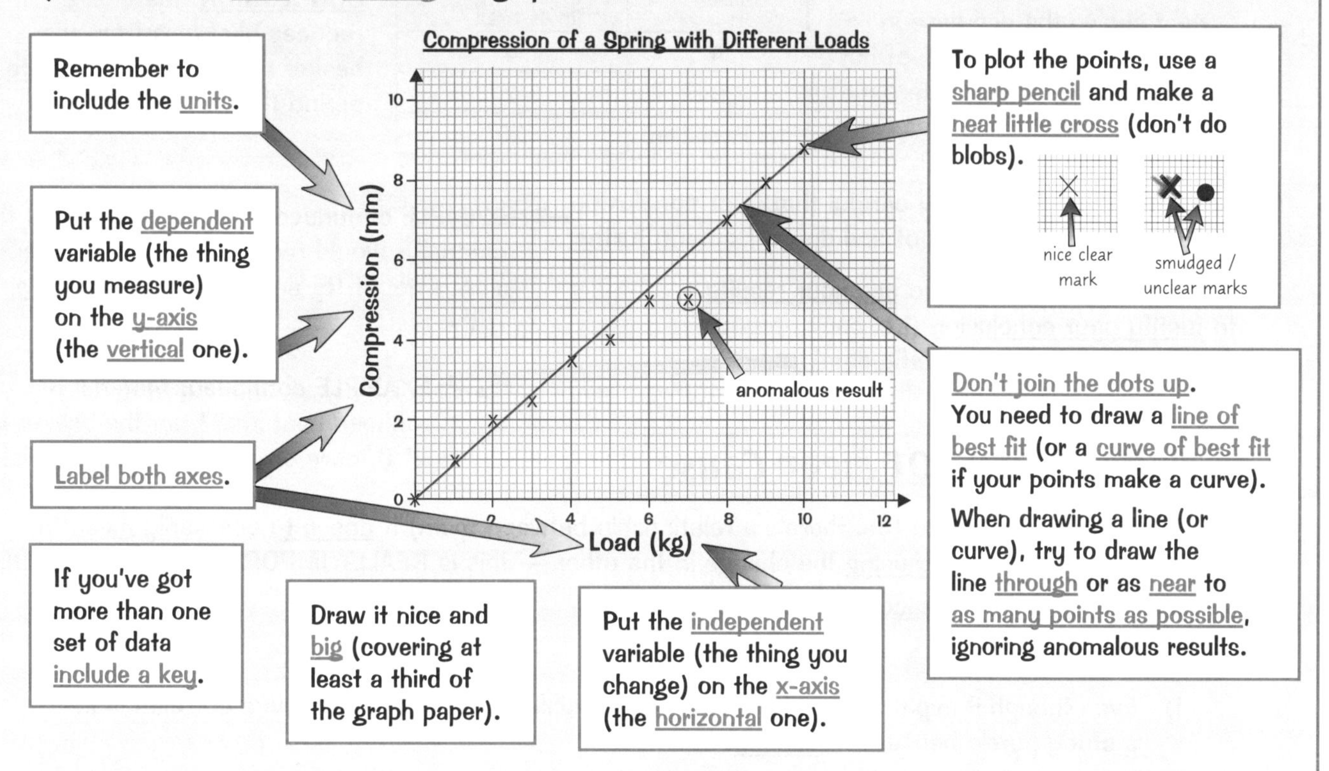

3) Line graphs are used to show the relationship between two variables (just like other graphs).
4) Data can show three different types of correlation (relationship):

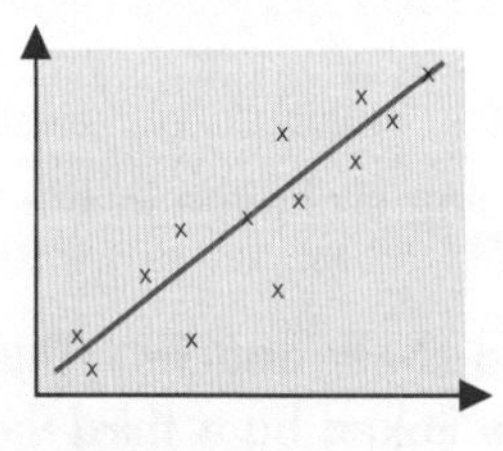

POSITIVE correlation — as one variable increases the other increases.

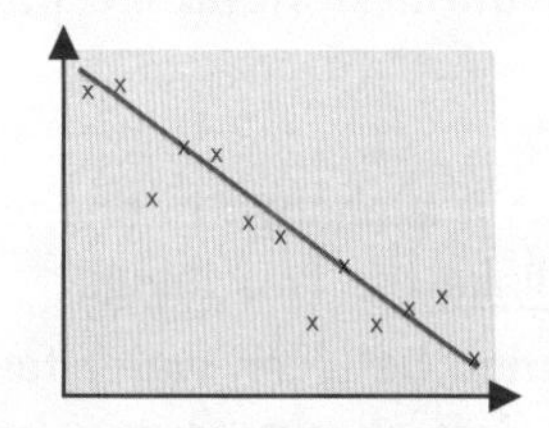

NEGATIVE correlation — as one variable increases the other decreases.

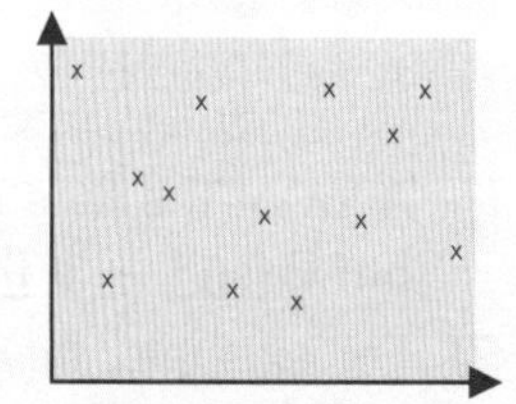

NO correlation — there's no relationship between the two variables.

5) You need to be able to describe the following relationships on line graphs too:

LINEAR — the graph is a straight line.

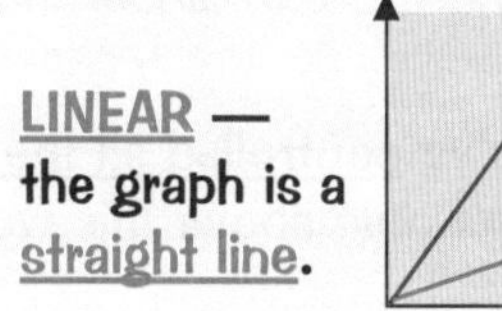

DIRECTLY PROPORTIONAL — the graph is a straight line where both variables increase (or decrease) in the same ratio.

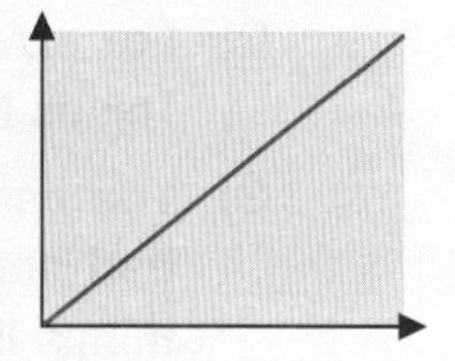

There's a positive correlation between revision and boredom...

...but there's also a positive correlation between revision and getting a better mark in the exam. Cover the page and write down the eight things you need to remember when drawing graphs. No sneaky peeking either — I saw you.

Drawing Conclusions

Congratulations — you've made it to the final step of a gruelling investigation — drawing conclusions.

You Can Only Conclude What the Data Shows and NO MORE

1) Drawing conclusions might seem pretty straightforward — you just look at your data and say what pattern or relationship you see between the dependent and independent variables.

EXAMPLE: The table on the right shows the decrease in temperature of a beaker of hot water insulated with different materials over 10 minutes.

Material	Mean temperature decrease (°C)
A	4
B	2
No insulation	20

CONCLUSION: Material B reduces heat loss from the beaker more over a 10 minute period than material A.

2) But you've got to be really careful that your conclusion matches the data you've got and doesn't go any further.

EXAMPLE continued: You can't conclude that material B would reduce heat loss by the same amount for any other type of container — the results could be totally different.

3) You also need to be able to use your results to justify your conclusion (i.e. back up your conclusion with some specific data).

EXAMPLE continued: Material B reduced heat loss from the beaker by 2 °C more on average than material A.

Correlation DOES NOT mean Cause

1) If two things are correlated (i.e. there's a relationship between them) it doesn't necessarily mean that a change in one variable is causing the change in the other — this is REALLY IMPORTANT, DON'T FORGET IT.
2) There are three possible reasons for a correlation:

1 CHANCE

1) Even though it might seem a bit weird, it's possible that two things show a correlation in a study purely because of chance.
2) For example, one study might find a correlation between people's hair colour and how good they are at frisbee. But other scientists don't get a correlation when they investigate it — the results of the first study are just a fluke.

2 LINKED BY A 3rd VARIABLE

1) A lot of the time it may look as if a change in one variable is causing a change in the other, but it isn't — a third variable links the two things.
2) For example, there's a correlation between water temperature and shark attacks. This obviously isn't because warmer water makes sharks crazy. Instead, they're linked by a third variable — the number of people swimming (more people swim when the water's hotter, and with more people in the water you get more shark attacks).

3 CAUSE

1) Sometimes a change in one variable does cause a change in the other.
2) For example, there's a correlation between exposure to radiation and thyroid cancer. This is because radiation can cause cancer.
3) You can only conclude that a correlation is due to cause when you've controlled all the variables that could, just could, be affecting the result. (For the radiation example above this would include things like age and exposure to other things that cause cancer).

I conclude that this page is a bit dull...

...although, just because I find it dull doesn't mean that I can conclude it's dull (you might think it's the most interesting thing since that kid got his head stuck in the railings near school). In the exams you could be given a conclusion and asked whether some data supports it — so make sure you understand how far conclusions can go.

Controlled Assessment (ISA)

Controlled Assessment involves doing an experiment and answering two question papers on it under exam conditions. Sounds thrilling.

There are Two Sections in the Controlled Assessment

1) Planning

Before you do the Section 1 question paper you'll be given time to do some research into the topic that's been set — you'll need to develop a hypothesis/prediction and come up with two different methods to test it. In your research, you should use a variety of different sources (e.g. the internet, textbooks etc.). You'll need to be able to outline both methods and say which one is best (and why it's the best one) and describe your preferred method in detail. You're allowed to write notes about your two methods on one side of A4 and have them with you for both question papers. In Section 1, you could be asked things like:

1) What your hypothesis/prediction is.
2) What variables you're going to control (and how you're going to control them).
3) What measurements you're going to take.
4) What range and interval of values you will use for the independent variable.
5) How you'd figure out the range and interval using a trial run (sometimes called a 'preliminary investigation' in the question papers). See page 6 for more.
6) How many times you're going to repeat the experiment — a minimum of three is a good idea.
7) What equipment you're going to use (and why that equipment is right for the job).
8) How to carry out the experiment, i.e. what you do first, what you do second...
9) What hazards are involved in doing the experiment, and how to reduce them.
10) What table you'll draw to put your results in. See page 8 for how to draw one that examiners will love.

There's lots of help on all of these things on pages 5-10.

When you've done the planning and completed the first question paper you'll actually do the experiment. Then you'll have to present your data. Make sure you use the right type of graph, and you draw it properly — see pages 8-9 for help. After that it's onto the Section 2 question paper...

2) Drawing Conclusions and Evaluating

For the Section 2 question paper you have to do these things for your experiment:

1) Analyse and draw conclusions from your results. For this you need to describe the relationship between the variables in detail — see the previous page for how to do this. E.g. 'I found that there is a relationship between picking your nose and having spots. The more often you pick your nose the more spots you'll have. For example, my results showed...'.
2) Say whether your results back up the hypothesis/prediction, and give reasons why or why not. E.g. 'My results did not back up the prediction. The prediction was that picking your nose more has no effect on the number of spots you have. But I found the opposite to be true in my investigation'.
3) Evaluate your experiment. For this you need to suggest ways you could improve your experiment.
   - Comment on your equipment and method, e.g. could you have used more accurate equipment?
   - Make sure you explain how the improvements would give you better data next time.
   - Refer to your results. E.g. 'My data wasn't accurate enough because the mass balance I used only measured in 1 g steps. I could use a more sensitive one next time (e.g. a mass balance that measures in 0.5 g steps) to get more accurate data'.

You'll also be given some secondary data (data collected by someone else) from an experiment on the same topic and asked to analyse it. This just involves doing what you did for your data with the secondary data, e.g. draw conclusions from it.

If that's controlled assessment, I'd hate to see uncontrolled assessment...

That might be an Everest-sized list of stuff, but it's all important. No need to panic at the sight of it though — as long as you've learnt everything on the previous few pages, you should be fine.

Heat Radiation

Heat energy tends to flow away from a hotter object to its cooler surroundings.
But then you knew that already. I would hope.

Heat Is Transferred in Three Different Ways

1) Heat energy can be transferred by radiation, conduction or convection.
2) Heat radiation is the transfer of heat energy by infrared (IR) radiation (see below).
3) Conduction and convection involve the transfer of energy by particles.
4) Conduction is the main form of heat transfer in solids (see p.13).
5) Convection is the main form of heat transfer in liquids and gases (see p.14).
6) Infrared radiation can be emitted by solids, liquids and gases.
7) Any object can both absorb and emit infrared radiation, whether or not conduction or convection are also taking place.
8) The bigger the temperature difference between a body and its surroundings, the faster energy is transferred by heating. Kinda makes sense.

Infrared Radiation — Emission of Electromagnetic Waves

1) All objects are continually emitting and absorbing infrared radiation. Infrared radiation is emitted from the surface of an object.
2) An object that's hotter than its surroundings emits more radiation than it absorbs (as it cools down). And an object that's cooler than its surroundings absorbs more radiation than it emits (as it warms up).
3) The hotter an object is, the more radiation it radiates in a given time.
4) You can feel this infrared radiation if you stand near something hot like a fire or if you put your hand just above the bonnet of a recently parked car.

(recently parked car)

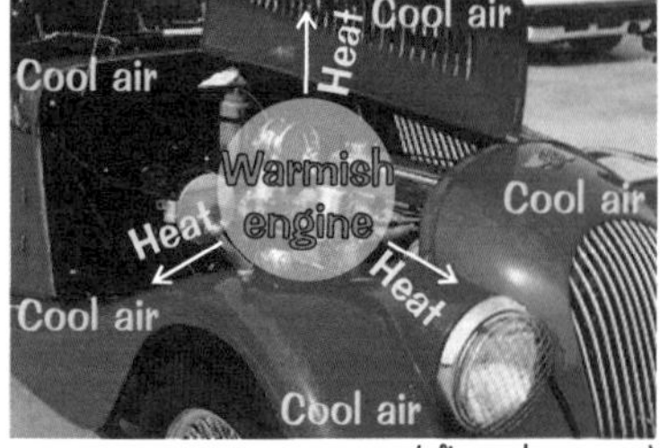

(after an hour or so)

Radiation Depends an Awful Lot on Surface Colour and Texture

1) Dark, matt surfaces absorb infrared radiation falling on them much better than light, shiny surfaces, such as gloss white or silver. They also emit much more infrared radiation (at any given temperature).
2) Light, shiny surfaces reflect a lot of the infrared radiation falling on them. E.g. vacuum flasks (see p.16) have silver inner surfaces to keep heat in or out, depending on whether it's storing hot or cold liquid.

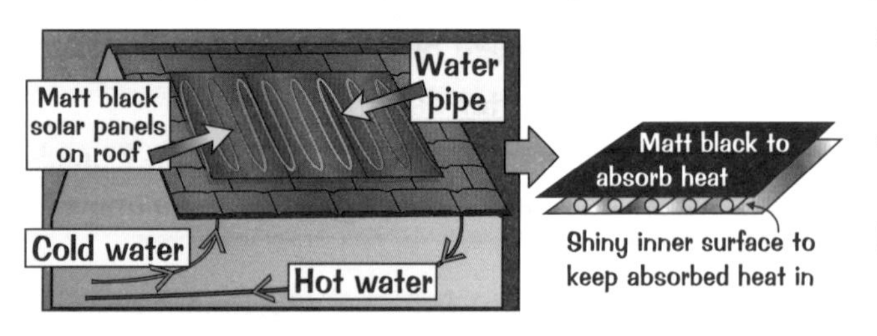

3) Solar hot water panels contain water pipes under a black surface (or black painted pipes under glass).
4) Radiation from the Sun is absorbed by the black surface to heat the water in the pipes.
5) This water can be used for washing or pumped to radiators to heat the building.

Feelin' hot hot hot...

You might be asked about an example of IR radiation that you've not come across before. As long as you remember that light, shiny surfaces reflect IR radiation and dark, matt surfaces absorb it — you should be able to figure out what's going on. If this stuff on radiation is floating your boat, you're going to love conduction...

Kinetic Theory and Conduction

Kinetic theory sounds complicated but it's actually pretty simple. It just describes how particles move in solids, liquids and gases. The energy an object (or particle) has because of its movement is called its kinetic energy.

Kinetic Theory Can Explain the Three States of Matter

The three states of matter are solid (e.g. ice), liquid (e.g. water) and gas (e.g. water vapour). The particles of a particular substance in each state are the same — only the arrangement and energy of the particles are different.

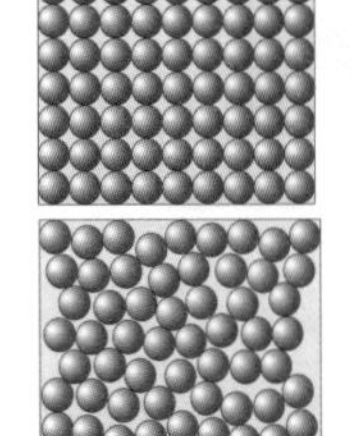

SOLIDS — strong forces of attraction hold the particles close together in a fixed, regular arrangement. The particles don't have much energy so they can only vibrate about their fixed positions.

LIQUIDS — there are weaker forces of attraction between the particles. The particles are close together, but can move past each other, and form irregular arrangements. They have more energy than the particles in a solid — they move in random directions at low speeds.

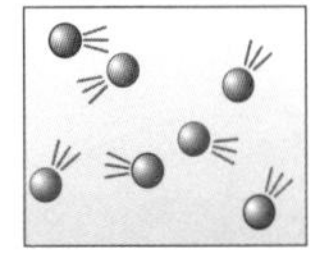

GASES — There are almost no forces of attraction between the particles. The particles have more energy than those in liquids and solids — they are free to move, and travel in random directions and at high speeds.

When you heat a substance, you give its particles more kinetic energy (KE) — they vibrate or move faster. This is what eventually causes solids to melt and liquids to boil.

Conduction of Heat — Occurs Mainly in Solids

OW!

Heat carries on conducting through metal pan handle

Heat conducts through pan to water

CONDUCTION OF HEAT ENERGY is the process where VIBRATING PARTICLES pass on their EXTRA KINETIC ENERGY to NEIGHBOURING PARTICLES.

This process continues throughout the solid and gradually some of the extra kinetic energy (or heat) is passed all the way through the solid, causing a rise in temperature at the other side of the solid. And hence an increase in the heat radiating from its surface.

Usually conduction is faster in denser solids, because the particles are closer together and so will collide more often and pass energy between them. Materials that have larger spaces between their particles conduct heat energy much more slowly — these materials are insulators.

Metals are Good Conductors Because of Their Free Electrons

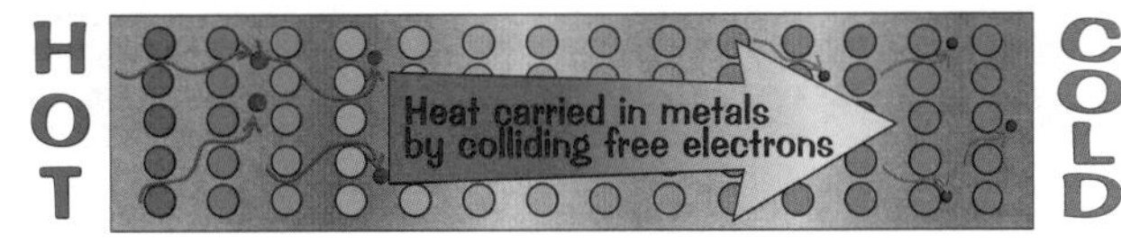

Conduction is more efficient through a short, fat rod than through a long, thin rod. It all comes down to how far the electrons have to transfer the energy.

1) Metals "conduct" so well because the electrons are free to move inside the metal.
2) At the hot end the electrons move faster and collide with other free electrons, transferring energy. These other electrons then pass on their extra energy to other electrons, etc.
3) Because the electrons can move freely, this is obviously a much faster way of transferring the energy through the metal than slowly passing it between jostling neighbouring atoms.
4) This is why heat energy travels so fast through metals.

Good conductors are always metals? — what about Henry Wood...

You'll notice that if a spade has been left in the sun for a while, the metal part will always feel much hotter than the wooden handle. But IT ISN'T HOTTER — it just conducts the heat into your hand much quicker than the wood, so your hand heats up much quicker. In cold weather, the metal bits of a spade, or anything else, always feel colder because they take the heat away from your hand quicker. But they're NOT COLDER... Remember that.

Convection

Gases and liquids are usually free to slosh about — and that allows them to transfer heat by convection, which is a much more effective process than conduction.

Convection of Heat — Liquids and Gases Only

CONVECTION occurs when the more energetic particles MOVE from the HOTTER REGION to the COOLER REGION — AND TAKE THEIR HEAT ENERGY WITH THEM.

This is how immersion heaters in kettles and hot water tanks and (unsurprisingly) convector heaters work. Convection simply can't happen in solids because the particles can't move.

The Immersion Heater Example

In a bit more detail:

1) Heat energy is transferred from the heater coils to the water by conduction (particle collisions).
2) The particles near the coils get more energy, so they start moving around faster.
3) This means there's more distance between them, i.e. the water expands and becomes less dense.
4) This reduction in density means that the hotter water tends to rise above the denser, cooler water.
5) As the hot water rises it displaces (moves) the colder water out of the way, making it sink towards the heater coils.
6) This cold water is then heated by the coils and rises — and so it goes on. You end up with convection currents going up, round and down, circulating the heat energy through the water.

Fast-moving particles collide with slow-moving particles & transfer heat

Less dense water rises

Hot water less dense

Water heats

Water cools and becomes more dense

Denser water sinks again

Water (and heat) circulates by convection

Heater coils

Almost no conduction

Water stays cold below the heater

Note that convection is most efficient in roundish or squarish containers, because they allow the convection currents to work best. Shallow, wide containers or tall, thin ones just don't work quite so well.

Also note that because the hot water rises (because of the lower density) you only get convection currents in the water above the heater. The water below it stays cold because there's almost no conduction.

CONVECTION CURRENTS are all about CHANGES IN DENSITY. Remember that.

The Radiator Example

1) Heating a room with a radiator relies on convection currents too.
2) Hot, less dense air by the radiator rises and denser, cooler air flows to replace it.

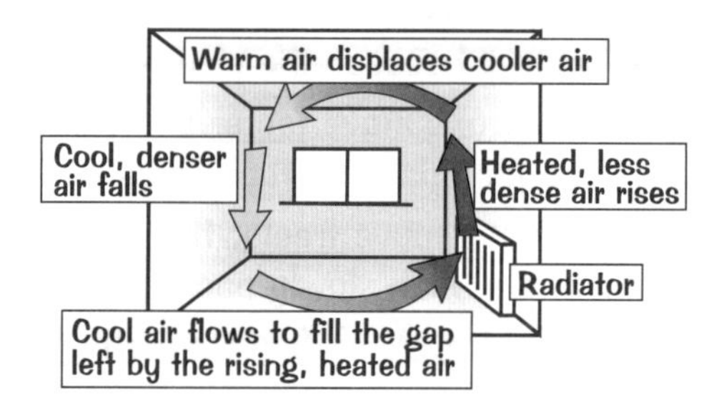

There's a great experiment with purple crystals to show this...

You stick some potassium permanganate crystals in the bottom of a beaker of cold water, then heat it gently over a Bunsen flame. The potassium permanganate starts to dissolve and make a gorgeous bright purple solution that gets moved around the beaker by the convection currents as the water heats. It's real pretty. ☺

Condensation and Evaporation

Here are a couple more things about particles in gases and liquids you need to think about. It's party-cle time...

Condensation is When Gas Turns to Liquid

1) When a gas cools, the particles in the gas slow down and lose kinetic energy. The attractive forces between the particles pull them closer together.
2) If the temperature gets cold enough and the gas particles get close enough together that condensation can take place, the gas becomes a liquid.
3) Water vapour in the air condenses when it comes into contact with cold surfaces e.g. drinks glasses.
4) The steam you see rising from a boiling kettle is actually invisible water vapour condensing to form tiny water droplets as it spreads into cooler air.

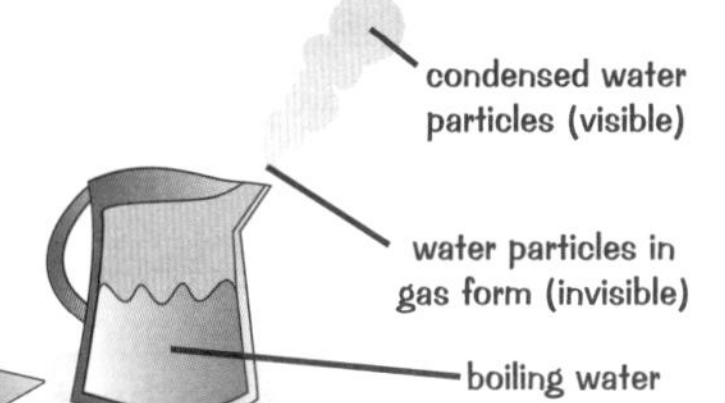

Evaporation is When Liquid Turns to Gas

1) Evaporation is when particles escape from a liquid.
2) Particles can evaporate from a liquid at temperatures that are much lower than the liquid's boiling point.
3) Particles near the surface of a liquid can escape and become gas particles if:

- The particles are travelling in the right direction to escape the liquid.
- The particles are travelling fast enough (they have enough kinetic energy) to overcome the attractive forces of the other particles in the liquid.

4) The fastest particles (with the most kinetic energy) are most likely to evaporate from the liquid — so when they do, the average speed and kinetic energy of the remaining particles decreases.
5) This decrease in average particle energy means the temperature of the remaining liquid falls — the liquid cools.
6) This cooling effect can be really useful. For example, you sweat when you exercise or get hot. As the water from the sweat on your skin evaporates, it cools you down.

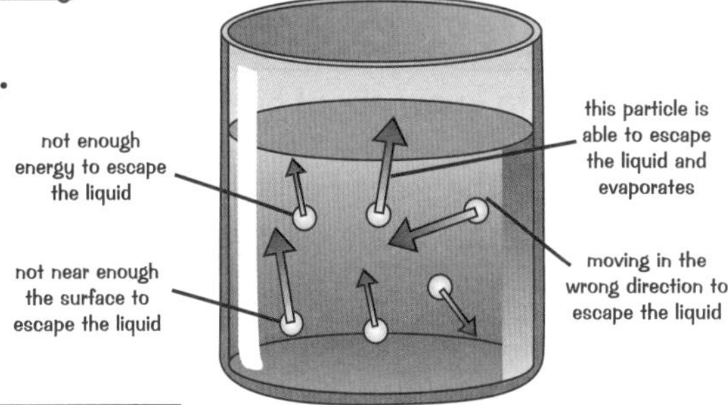

Rates of Evaporation and Condensation can Vary

The RATE OF EVAPORATION will be faster if the...

- TEMPERATURE is higher — the average particle energy will be higher, so more particles will have enough energy to escape.
- DENSITY is lower — the forces between the particles will usually be weaker, so more particles will have enough energy to overcome these forces and escape the liquid.
- SURFACE AREA is larger — more particles will be near enough to the surface to escape the liquid.
- AIRFLOW over the liquid is greater — the lower the concentration of an evaporating substance in the air it's evaporating into, the higher the rate of evaporation. A greater airflow means air above the liquid is replaced more quickly, so the concentration in the air will be lower.

The RATE OF CONDENSATION will be faster if the...

- TEMPERATURE OF THE GAS is lower — the average particle energy in the gas is lower — so more particles will slow down enough to clump together and form liquid droplets.
- TEMPERATURE OF THE SURFACE THE GAS TOUCHES is lower.
- DENSITY is higher — the forces between the particles will be stronger. Fewer particles will have enough energy to overcome these forces and will instead clump together and form a liquid.
- AIRFLOW is less — the concentration of the substance in the air will be higher, and so the rate of condensation will be greater.

A little less condensation, a little more action...

The people who make adverts for drinks know what customers like to see — condensation on the outside of the bottle. It makes the drink look nice and cold and extra-refreshing. Mmmm. If it wasn't for condensation, you'd never be able to draw pictures on the bus window with your finger either — you've got a lot to be thankful for...

Rate of Heat Transfer

There are loads of factors that affect the rate of heat transfer.
Different objects can lose or gain heat much faster than others — even in the same conditions. Read on...

The Rate of Heat Energy Transfer Depends on Many Things...

1) Heat energy is radiated from the surface of an object.
2) The bigger the surface area, the more infrared waves that can be emitted from (or absorbed by) the surface — so the quicker the transfer of heat. E.g. radiators have large surface areas to maximise the amount of heat they transfer.
3) This is why car and motorbike engines often have 'fins' — they increase the surface area so heat is radiated away quicker. So the engine cools quicker.
4) Heat sinks are devices designed to transfer heat away from objects they're in contact with, e.g. computer components. They have fins and a large surface area so they can emit heat as quickly as possible.

Cooling fins on engines increase surface area to speed up cooling.

5) If two objects at the same temperature have the same surface area but different volumes, the object with the smaller volume will cool more quickly — as a higher proportion of the object will be in contact with its surroundings.
6) Other factors, like the type of material, affect the rate too. Objects made from good conductors (see p.13) transfer heat away more quickly than insulating materials, e.g. plastic. It also matters whether the materials in contact with it are insulators or conductors. If an object is in contact with a conductor, the heat will be conducted away much faster than if it is in contact with a good insulator.

Some Devices are Designed to Limit Heat Transfer

You need to know about heat energy transfers and how products can be designed to reduce them.

Vacuum Flasks

1) The glass bottle is double-walled with a vacuum between the two walls. This stops all conduction and convection through the sides.
2) The walls either side of the vacuum are silvered to keep heat loss by radiation to a minimum.
3) The bottle is supported using insulating foam. This minimises heat conduction to or from the outer glass bottle.
4) The stopper is made of plastic and filled with cork or foam to reduce any heat conduction through it.

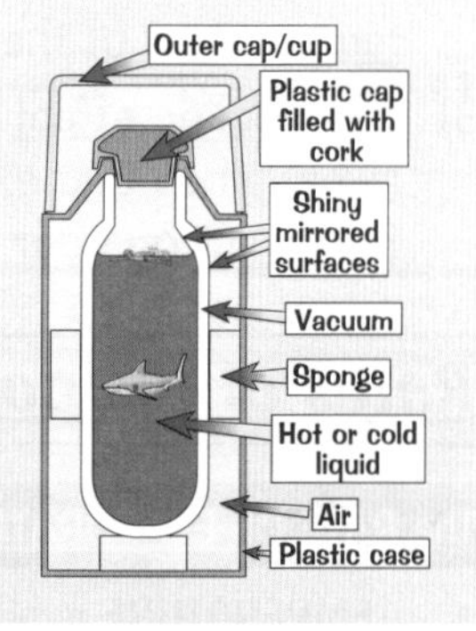

Humans and Animals Have Ways of Controlling Heat Transfer Too

1) In the cold, the hairs on your skin 'stand up' to trap a thicker layer of insulating air around the body. This limits the amount of heat loss by convection. Some animals do the same using fur.
2) When you're too warm, your body diverts more blood to flow near the surface of your skin so that more heat can be lost by radiation — that's why some people go pink when they get hot.
3) Generally, animals in warm climates have larger ears than those in cold climates to help control heat transfer.

For example, Arctic foxes have evolved small ears, with a small surface area to minimise heat loss by radiation and conserve body heat.
Desert foxes on the other hand have huge ears with a large surface area to allow them to lose heat by radiation easily and keep cool.

Don't call me 'Big Ears' — call me 'Large Surface Area'...

Examiners are like small children — they ask some barmy questions. If they ask you one about heat transfer, you must always say which form of heat transfer is involved at any point, either conduction, convection or radiation. You've got to show them that you know your stuff — it's the only way to get top marks.

Energy Efficiency in the Home

There are lots of things you can do to a building to reduce the amount of heat energy that escapes. Some are more effective than others, and some are better for your pocket than others. The most obvious examples are in the home, but you could apply this to any situation where you're trying to cut down energy loss.

Effectiveness and Cost-effectiveness are Not the Same...

Loft Insulation
Initial Cost: £200
Annual Saving: £50
Payback time: 4 years

Hot Water Tank Jacket
Initial Cost: £15
Annual Saving: £30
Payback time: 6 months

Double Glazing
Initial Cost: £3000
Annual Saving: £60
Payback time: 50 years

Cavity Wall Insulation
Initial Cost: £500
Annual Saving: £70
Payback time: 7 years

Draught-proofing
Initial Cost: £100
Annual Saving: £50
Payback time: 2 years

1) The most effective methods of insulation are ones that give you the biggest annual saving (they save you the most money each year on your heating bills).
2) Eventually, the money you've saved on heating bills will equal the initial cost of putting in the insulation (the amount it cost to buy). The time it takes is called the payback time.
3) The most cost-effective methods tend to be the cheapest.
4) They are cost-effective because they have a short payback time — this means the money you save covers the amount you paid really quickly.

Know Which Types of Heat Transfer Are Involved

1) CAVITY WALL INSULATION — foam squirted into the gap between the bricks reduces convection and radiation across the gap.
2) LOFT INSULATION — a thick layer of fibreglass wool laid out across the whole loft floor reduces conduction and radiation into the roof space from the ceiling.
3) DRAUGHT-PROOFING — strips of foam and plastic around doors and windows stop draughts of cold air blowing in, i.e. they reduce heat loss due to convection.
4) HOT WATER TANK JACKET — lagging such as fibreglass wool reduces conduction and radiation.
5) THICK CURTAINS — big bits of cloth over the window to reduce heat loss by conduction and radiation.

U-Values Show How Fast Heat can Transfer Through a Material

1) Heat transfers faster through materials with higher U-values than through materials with low U-values.
2) So the better the insulator (see p.13) the lower the U-value. E.g. The U-value of a typical duvet is about 0.75 W/m^{2}K, whereas the U-value of loft insulation material is around 0.15 W/m^{2}K.

It's payback time...

And it's the same with, say, cars. Buying a more fuel-efficient car might sound like a great idea — but if it costs loads more than a clapped-out old fuel-guzzler, you might still end up out of pocket. If it's cost-effectiveness you're thinking about, you always have to offset initial cost against annual savings.

Specific Heat Capacity

Specific heat capacity is one of those topics that puts people off just because it has a weird name. If you can get over that, it's actually not too bad — it sounds a lot harder than it is. Go on. Give it a second chance.

Specific Heat Capacity Tells You How Much Energy Stuff Can Store

1) It takes more heat energy to increase the temperature of some materials than others. E.g. you need 4200 J to warm 1 kg of water by 1 °C, but only 139 J to warm 1 kg of mercury by 1 °C.
2) Materials which need to gain lots of energy to warm up also release loads of energy when they cool down again. They can 'store' a lot of heat.
3) The measure of how much energy a substance can store is called its specific heat capacity.
4) Specific heat capacity is the amount of energy needed to raise the temperature of 1 kg of a substance by 1 °C. Water has a specific heat capacity of 4200 J/kg°C.

There's a Handy Formula for Specific Heat Capacity

You'll have to do calculations involving specific heat capacity. This is the equation to learn:

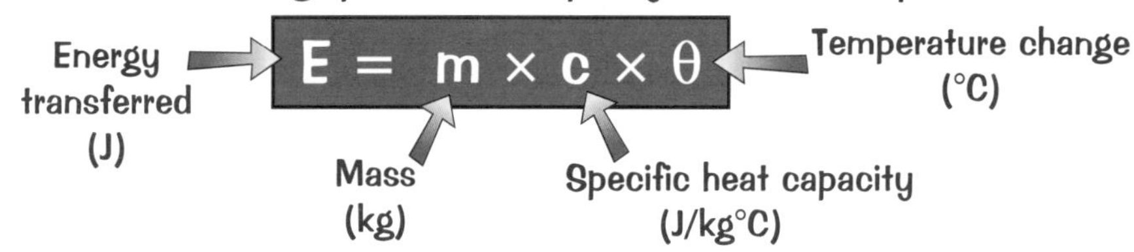

EXAMPLE: How much energy is needed to heat 2 kg of water from 10 °C to 100 °C?

ANSWER: Energy needed = $2 \times 4200 \times 90$ = 756 000 J

If you're not working out the energy, you'll have to rearrange the equation, so this formula triangle will come in dead handy. You cover up the thing you're trying to find. The parts of the formula you can still see are what it's equal to.

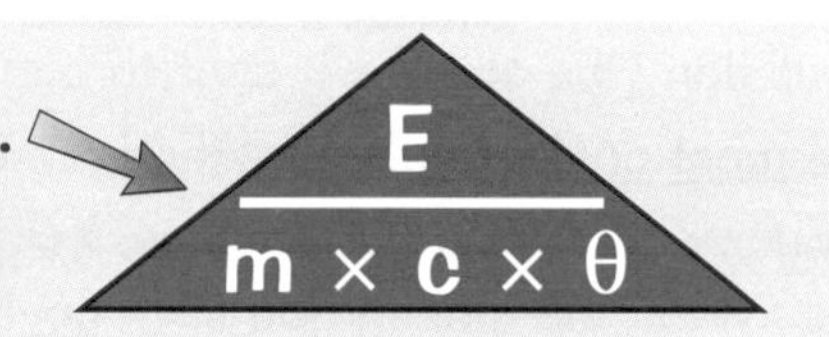

EXAMPLE: An empty 200 g aluminium kettle cools down from 115 °C to 10 °C, losing 19 068 J of heat energy. What is the specific heat capacity of aluminium?

Remember — you need to convert the mass to kilograms first.

ANSWER: $\text{SHC} = \frac{\text{Energy}}{\text{Mass} \times \text{Temp Ch}} = \frac{19\,068}{0.2 \times 105}$ = 908 J/kg°C

Heaters Have High Heat Capacities to Store Lots of Energy

1) The materials used in heaters usually have high specific heat capacities so that they can store large amounts of heat energy.
2) Water has a really high specific heat capacity. It's also a liquid, so it can easily be pumped around in pipes — ideal for central heating systems in buildings.
3) Electric storage heaters are designed to store heat energy at night (when electricity is cheaper), and then release it during the day. They store the heat using concrete or bricks, which (surprise surprise) have a high specific heat capacity (around 880 J/kg°C).
4) Some heaters are filled with oil, which has a specific heat capacity of around 2000 J/kg°C. Because this is lower than water's specific heat capacity, oil heating systems are often not as good as water-based systems. Oil does have a higher boiling point though, which usually means oil-filled heaters can safely reach higher temperatures than water-based ones.

I've just eaten five sausages — I have a high specific meat capacity...

I'm sure you'll agree that this isn't the most exciting part of GCSE physics — it's not about space travel, crashing cars or even using springs — but it is likely to come up in your GCSEs. Sadly you just have to knuckle down and get that formula triangle learnt — then you'll be well on the way to breezing through this question in the exam.

Energy Transfer

Heat is just one type of energy, but there are lots more as well:

Learn These Nine Types of Energy

You should know all of these well enough by now to list them from memory, including the examples:

1) ELECTRICAL Energy.................................... — whenever a current flows.
2) LIGHT Energy.. — from the Sun, light bulbs, etc.
3) SOUND Energy.. — from loudspeakers or anything noisy.
4) KINETIC Energy, or MOVEMENT Energy......... — anything that's moving has it.
5) NUCLEAR Energy...................................... — released only from nuclear reactions.
6) THERMAL Energy or HEAT Energy............... — flows from hot objects to colder ones.
7) GRAVITATIONAL POTENTIAL Energy.............. — possessed by anything which can fall.
8) ELASTIC POTENTIAL Energy......................... — stretched springs, elastic, rubber bands, etc.
9) CHEMICAL Energy...................................... — possessed by foods, fuels, batteries etc.

Potential- and Chemical-Energy Are Forms of Stored Energy

The last three above are forms of stored energy because the energy is not obviously doing anything, it's kind of waiting to happen, i.e. waiting to be turned into one of the other forms.

You Need to Know the Conservation of Energy Principle

There are plenty of different types of energy, but they all obey the principle below:

ENERGY CAN BE TRANSFERRED USEFULLY FROM ONE FORM TO ANOTHER, STORED OR DISSIPATED — BUT IT CAN NEVER BE CREATED OR DESTROYED.

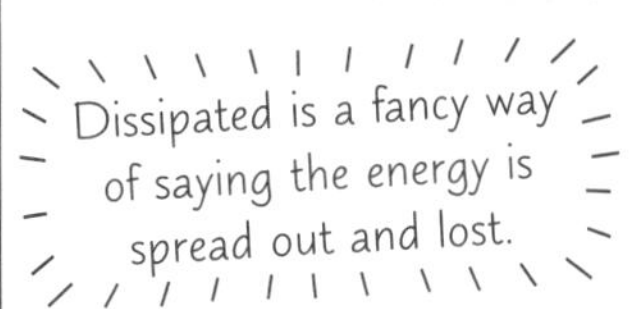

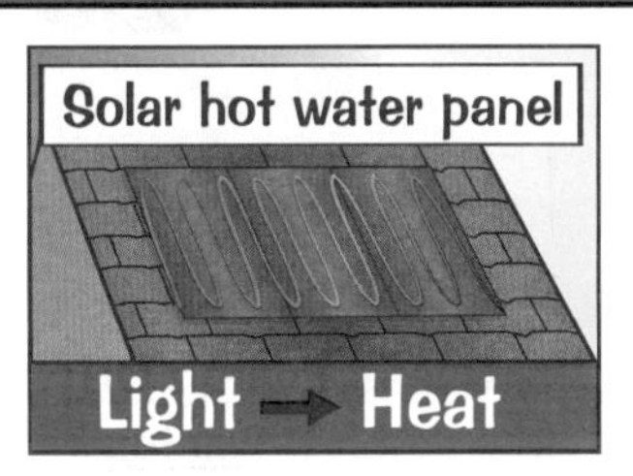

Gravitational Potential → Kinetic

Another important principle which you need to learn is this one:

Energy is only useful when it can be converted from one form to another.

They Like Giving Exam Questions on Energy Transfers

In the exam, they can ask you about any device or energy transfer system they feel like. If you understand a few different examples, it'll be easier to think through whatever they ask you about in the exam.

EXAMPLES:

Electrical Devices, e.g. televisions: Electrical energy → Light, sound and heat energy
Batteries: Chemical energy → Electrical and heat energy
Electrical Generation, e.g. wind turbines: Kinetic energy → Electrical and heat energy
Potential Energy, e.g. firing a bow and arrow: Elastic potential energy → Kinetic and heat energy

Energy can't be created or destroyed — only talked about a lot...

Chemical energy → kinetic energy → electrical energy → kinetic energy → chemical energy.
(me thinking) (me typing) (my computer) (printing machine) (you reading this)

Efficiency of Machines

More! More! Tell me more about energy transfers please! OK, since you insist:

Most Energy Transfers Involve Some Losses, Often as Heat

1) Useful devices are only useful because they can transform energy from one form to another.
2) In doing so, some of the useful input energy is always lost or wasted, often as heat.
3) The less energy that is 'wasted', the more efficient the device is said to be.
4) The energy flow diagram is pretty much the same for all devices.

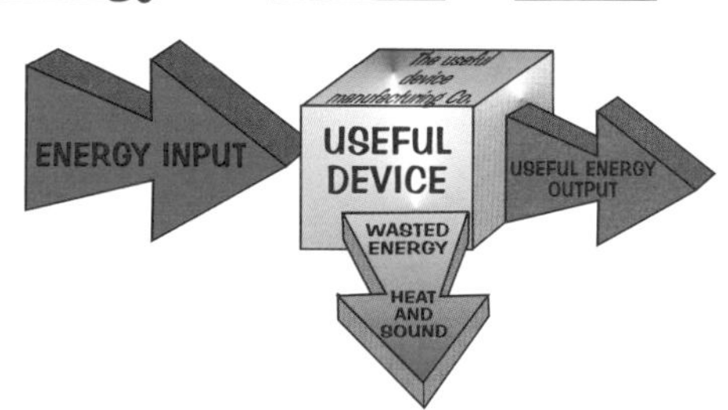

It's Really Simple to Calculate the Efficiency...

A machine is a device which turns one type of energy into another. The efficiency of any device is defined as:

$$\text{Efficiency} = \frac{\text{Useful Energy out}}{\text{Total Energy in}}$$

You might not know the energy inputs and outputs of a machine, but you can still calculate the machine's efficiency as long as you know the power input and output:

$$\text{Efficiency} = \frac{\text{Useful Power out}}{\text{Total Power in}}$$

You can give efficiency as a decimal or you can multiply your answer by 100 to get a percentage, i.e. 0.75 or 75%.

As usual, a formula triangle will come handy for rearranging the formulas:

$$\frac{\text{Useful Out}}{\text{Efficiency} \times \text{Total In}}$$

How to Use the Formula — Nothing to It

1) You find how much energy is supplied to a machine. (The Total Energy IN.)
2) You find how much useful energy the machine delivers. (The Useful Energy OUT.) An exam question either tells you this directly or tells you how much it wastes as heat/sound.
3) Either way, you get those two important numbers and then just divide the smaller one by the bigger one to get a value for efficiency somewhere between 0 and 1 (or 0 and 100%). Easy.
4) The other way they might ask it is to tell you the efficiency and the input energy and ask for the energy output — so you need to be able to swap the formula round.

Useful Energy Input Isn't Usually Equal to Total Energy Output

For any given example you can talk about the types of energy being input and output, but remember this:

No device is 100% efficient and the wasted energy is usually spread out as heat.

Electric heaters are the exception to this. They're usually 100% efficient because all the electricity is converted to "useful" heat. Ultimately, all energy ends up as heat energy. If you use an electric drill, it gives out various types of energy but they all quickly end up as heat.

Don't waste your energy — turn the TV off while you revise...

And for 10 bonus points, calculate the efficiency of these machines:

TV — energy in = 220 J, light energy out = 5 J, sound energy out = 2 J, heat energy out = 213 J.

Loudspeaker — energy in = 35 J, sound energy out = 0.5 J, heat energy out = 34.5 J. Answers on p.108.

Efficiency of Machines

I know what you're thinking — those inefficient machines are causing senseless waste. I'm pretty darn angry too. But sometimes efficiency isn't everything — there are other factors to consider too.

We Call It Wasted Heat Because We Can't Do Anything Useful with It

1) Useful energy is concentrated energy. As you know, the entire energy output by a machine, both useful and wasted, eventually ends up as heat.
2) This heat is transferred to cooler surroundings, which then become warmer. As the heat is transferred to cooler surroundings, the energy becomes less concentrated — it dissipates.
3) The total amount of energy stays the same. The energy is still there, but as it becomes increasingly spread out, it can't be easily used or collected back in again.

You Need to Think About Cost-Effectiveness and Efficiency... ...When Choosing Appliances

Example: Light Bulbs

1) A low-energy bulb is about 4 times as efficient as an ordinary light bulb.
2) Energy-efficient light bulbs are more expensive to buy but they last much longer.
3) If an energy-saving light bulb cost £3 and saved £12 of energy a year, its payback time (see p.17) would be 3 months.

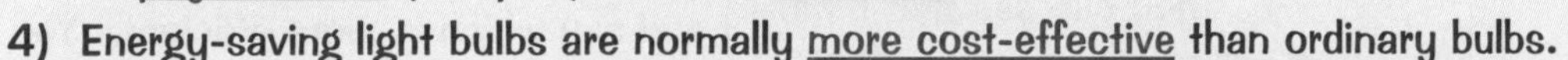

4) Energy-saving light bulbs are normally more cost-effective than ordinary bulbs.
5) LED light bulbs are even more efficient than low-energy bulbs, and can last even longer.
6) But they are more expensive to buy and don't give out as much light as the other two types of bulb.

Example: Replacing Old Appliances with Newer Energy-Efficient Ones

1) New, efficient appliances are cheaper to run than older, less efficient appliances. But new appliances can be expensive to buy.
2) You've got to work out if it's cost-effective (p.17) to buy a new appliance.
3) To work out how cost-effective a new appliance will be you need to work out its payback time.

Sometimes 'Waste' Energy Can Actually Be Useful

1) Heat exchangers reduce the amount of heat energy that is 'lost'.
2) They do this by pumping a cool fluid through the escaping heat.
3) The temperature of this fluid rises as it gains heat energy.
4) The heat energy in the fluid can then be converted into a form of energy that's useful again — either in the original device, or for other useful functions. For example, some of the heat from a car's engine can be transferred to the air that's used to warm the passenger compartment.

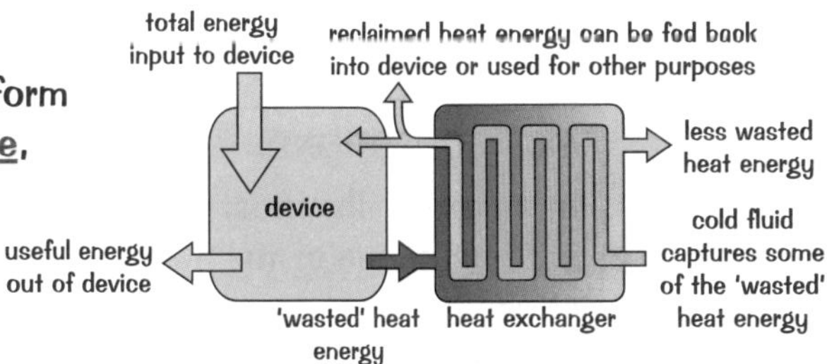

Let there be light — and a bit of wasted heat...

The thing about loss of energy is it's always the same — it always disappears as heat and sound, and even the sound ends up as heat pretty quickly. So when they ask, "Why is the input energy more than the output energy?", the answer is always the same... Learn and enjoy.

Energy Transformation Diagrams

This is another opportunity for a MATHS question. Fantastic.
So best prepare yourself — here's what those energy transformation diagrams are all about...

The Thickness of the Arrow Represents the Amount of Energy

The idea of Sankey diagrams is to make it easy to see at a glance how much of the total energy in is being usefully employed compared with how much is being wasted.

The thicker the arrow, the more energy it represents — so you see a big thick arrow going in, then several smaller arrows going off it to show the different energy transformations taking place.

You can have either a little sketch or a properly detailed diagram where the width of each arrow is proportional to the number of joules it represents.

Example — TV:

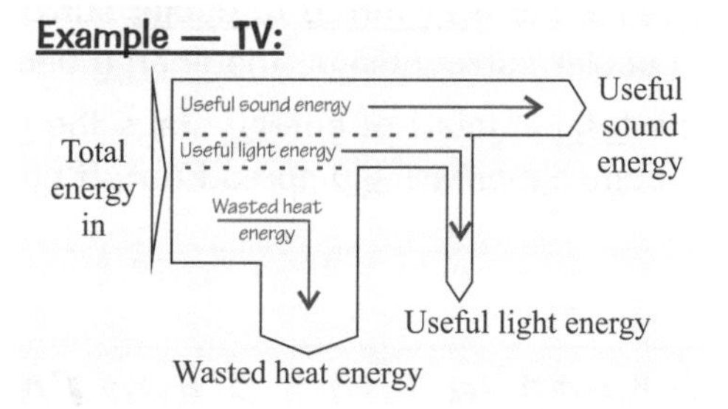

Example — Sankey Diagram for a Simple Motor:

HERE'S THE SKETCH VERSION:

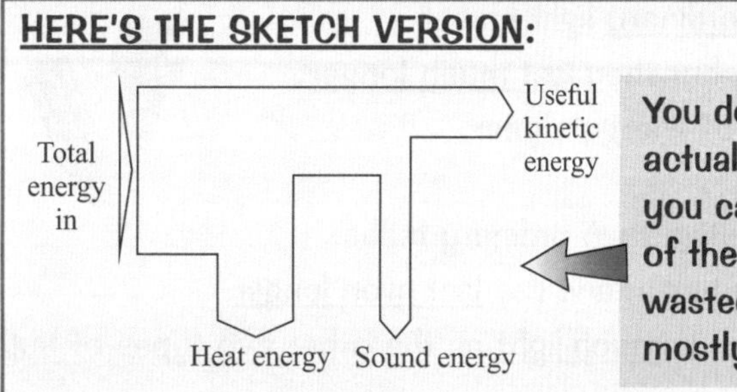

You don't know the actual amounts, but you can see that most of the energy is being wasted, and that it's mostly wasted as heat.

EXAM QUESTIONS:

With sketches, they're likely to ask you to compare two different devices and say which is more efficient. You generally want to be looking for the one with the thickest useful energy arrow(s).

AND HERE'S THE DETAILED ONE:

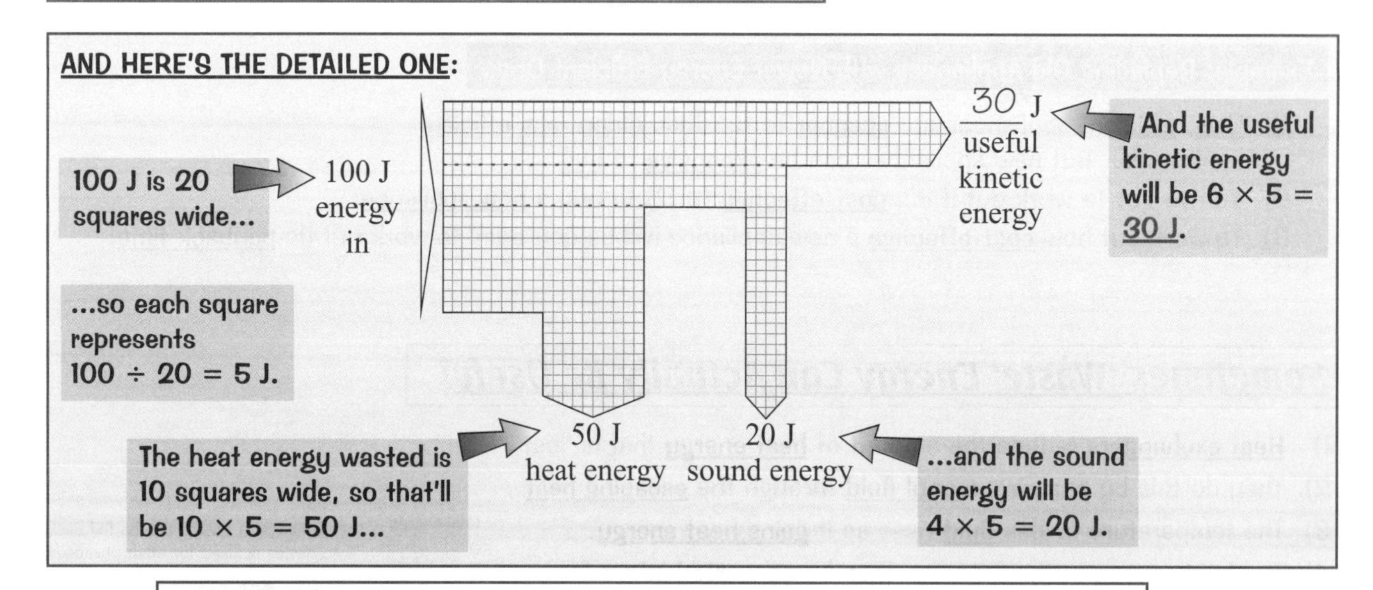

EXAM QUESTIONS:

In an exam, the most likely question you'll get about detailed Sankey diagrams is filling in one of the numbers or calculating the efficiency. The efficiency is straightforward enough if you can work out the numbers (see p.20).

Skankey diagrams — to represent the smelliness of your socks...

If they ask you to draw your own Sankey diagram in the exam, and don't give you the figures, a sketch is all they'll expect. Just give a rough idea of where the energy goes. E.g. a filament lamp turns most of the input energy into heat, and only a tiny proportion goes to useful light energy.

The Cost of Electricity

Isn't electricity great — generally, I mean. You can power all sorts of toys and gadgets with electricity. But it'll cost you. 'How much?' I hear you cry... Read and learn.

Kilowatt-hours (kWh) are "UNITS" of Energy

1) Electrical appliances transfer electrical energy into other forms (see page 19) — e.g. sound and heat energy in a radio.
2) The amount of energy that is transferred by an appliance depends on its power (how fast the appliance can transfer it) and the amount of time that the appliance is switched on.

ENERGY = POWER x TIME

3) Energy is usually measured in joules (J) — 1 J is the amount of energy transferred by a 1 W appliance in 1 s.
4) Power is usually measured in watts (W) or kilowatts (kW). A 5 kW appliance transfers 5000 J in 1 s.
5) When you're dealing with large amounts of electrical energy (e.g. the energy used by a home in one week), it's easier to think of the power and time in kilowatts and hours — rather than in watts and seconds.
6) So the standard units of electrical energy are kilowatt-hours (kWh) — not joules.

A KILOWATT-HOUR is the amount of electrical energy used by a 1 kW appliance left on for 1 HOUR.

The Two Easy Formulas for Calculating the Cost of Electricity

These must surely be the two most trivial and obvious formulas you'll ever see:

No. of UNITS (kWh) used = POWER (in kW) × TIME (in hours) — Units = kW × hours

COST = No. of UNITS × PRICE per UNIT — Cost = Units × Price

EXAMPLE: An electricity supplier charges 14p per unit.
Find the cost of leaving a 60 W light bulb on for: a) 30 minutes b) one year.

ANSWER: a) No. of units = kW × hours = 0.06 kW × ½ hr = 0.03 units.
Cost = units × price per unit(14p) = 0.03 × 14p = 0.42p for 30 mins.

b) No. of units = kW × hours = 0.06 kW × (24×365) hr = 525.6 units.
Cost = units × price per unit(14p) = 525.6 × 14p = £73.58 for one year.

EXAMPLE 2: Each unit of electricity costs 14p. For how long can a 6 kW heater be used for 14p?
A 6 hours B 1 hour C 10 minutes D 7 hours

ANSWER 2: The cost of 1 unit is 14p. So for 14p you can use 1 unit.
UNITS = POWER × TIME, so TIME = UNITS ÷ POWER = 1 ÷ 6 = 0.167 hours = 10 mins

You Need to Know How to Read an Electricity Meter

1) They might ask you to read values off an electricity meter in the exam — but don't worry, it's pretty straightforward. The units are usually in kWh — but make sure you check.
2) You could be given two meter readings and be asked to work out the total energy that's been used over a particular time period. Just subtract the meter reading at the start of the time (the smaller one) from the reading at the end to work this out.

500 kWh doesn't mean much to anyone — £70 is far more real...

In reality most electricity suppliers have complicated formulas for working out the cost of electricity. Luckily in the exam you'll just be told how much a particular energy supplier charges per unit or something, phew.

Choosing Electrical Appliances

Unfortunately, this isn't about what colour MP3 player to get, but know it you must I'm afraid...

Sometimes You Have a Choice of Electrical Equipment

1) There are often a few different appliances that do the same job. In the exam, they might ask you to weigh up the pros and cons of different appliances and decide which one is most suitable for a particular situation.
2) You might need to work out whether one appliance uses less energy or is more cost-effective than another.
3) You might also need to think about the practical advantages and disadvantages of using different appliances. E.g. 'Can an appliance be used in areas with limited electricity supplies?'
4) You might get asked to compare two appliances that you haven't seen before. Just take your time and think about the advantages and disadvantages — you should be able to make a sensible judgement.

Here are some examples of the sorts of things you might have to compare:

E.G. CLOCKWORK RADIOS AND BATTERY RADIOS

1) Battery radios and clockwork radios are both handy in areas where there is no mains electricity supply.
2) Clockwork radios work by storing elastic potential energy in a spring when someone winds them up. The elastic potential energy is slowly released and used to power the radio.
3) Batteries can be expensive, but powering a clockwork radio is free.
4) Battery power is also only useful if you can get hold of some new batteries when the old ones run out. You don't get that problem with clockwork radios — but it can get annoying to have to keep winding them up every few hours to recharge them.
5) Clockwork radios are also better for the environment — a lot of energy and harmful chemicals go into making batteries, and they're often tricky to dispose of safely.

You Might Be Asked to Use Data to Compare Two Appliances

EXAMPLE

A company is deciding whether to install a 720 W low-power heater, or a high-power 9 kW heater. The heater they choose will be on for 30 hours each week. Their electricity provider charges 7p per kWh of electricity. How much money per week would they save by choosing the low-power heater?

ANSWER: Weekly electricity used by the low-power heater = 0.720 kW × 30 h = 21.6 kWh
Weekly electricity used by the high-power heater = 9 kW × 30 h = 270 kWh
Total saving = (270 – 21.6) × 7 = £17.39 (to the nearest penny)

Standard of Living is Affected by Access to Electricity

1) Most people in developed countries have access to mains electricity. However, many people living in the world's poorest countries don't — this has a big effect on their standard of living.
2) In the UK, our houses are full of devices that transform electrical energy into other useful types of energy. For example, not only is electric lighting useful and convenient, but it can also help improve safety at night.
3) Refrigerators keep food fresh for longer by slowing down the growth of bacteria. Refrigerators are also used to keep vaccines cold. Without refrigeration it's difficult to distribute important vaccines — this can have devastating effects on a country's population.
4) Electricity also plays an important role in improving public health in other ways. Hospitals in developed countries rely heavily on electricity, e.g. for X-ray machines. Without access to these modern machines, the diagnosis and treatment of patients would be poorer and could reduce life expectancy.
5) Communications are also affected by a lack of electricity. No electricity means no internet or phones — making it hard for people to keep in touch, or for people to send and receive news and information.

I'm definitely a fan of things running like clockwork...

Make sure you're happy with comparing electrical devices, and you know how important access to electricity can be.

Revision Summary for Physics 1a

It's all very well reading the pages and looking at the diagrams — but you won't have a hope of remembering it for your exam if you don't understand it. Have a go at these questions to see how much has gone in so far. If you struggle with any of them, have another read through the section and give the questions another go.

1) Describe the three ways that heat energy can be transferred.
2) True or false? An object that's cooler than its surroundings emits more radiation than it absorbs.
3) Explain why solar hot water panels have a matt black surface.
4) Describe the arrangement and movement of the particles in a) solids b) liquids c) gases
5) What is the name of the process where vibrating particles pass on their extra kinetic energy to neighbouring particles?
6) Which type of heat transfer can't take place in solids — convection or conduction?
7) Describe how the heat from heater coils is transferred throughout the water in a kettle. What is this process called?
8) How do the densities of liquids and gases change as you heat them?
9) What happens to the particles of a gas as it turns to a liquid?
10) What is the name given to the process where a gas turns to a liquid?
11) Why does evaporation have a cooling effect on a liquid?
12) The two designs of car engine shown are made from the same material. Which engine will transfer heat quicker? Explain why.

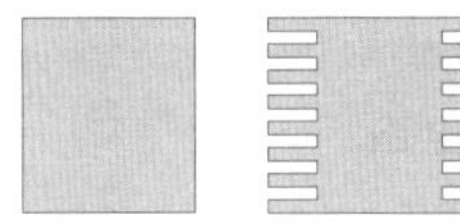

13) Describe two features of a vacuum flask that make it good at keeping hot liquids hot.
14) Do animals that live in hot climates tend to have large or small ears? Give one reason why this might be an advantage in a hot climate.
15)*If it costs £4000 to double glaze your house and the double glazing saves you £100 on energy bills every year, calculate the payback time for double glazing.
16) Name five ways of improving energy efficiency in the home. Explain how each improvement reduces the amount of heat lost from a house.
17) What can you tell from a material's U-value?
18) Would you expect copper or cotton wool to have a higher U-value?
19) What property of a material tells you how much energy it can store?
20)*An ornament has a mass of 0.5 kg. The ornament is made from a material that has a specific heat capacity of 1000 J/kg°C. How much energy does it take to heat the ornament from 20 °C to 200 °C?
21) Do heaters use materials that have a high or low heat capacity?
22) Name nine types of energy and give an example of each.
23) State the principle of the conservation of energy.
24) List the energy transformations that occur in a battery-powered toy car.
25) What is the useful type of energy delivered by a motor? In what form is energy wasted?
26)*What is the efficiency of a motor that converts 100 J of electrical energy into 70 J of useful kinetic energy?
27)*The following Sankey diagram shows how energy is converted in a catapult.

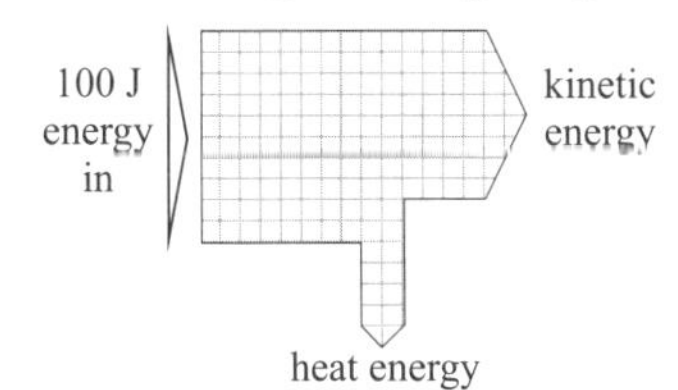

a) How much energy is converted into kinetic energy?
b) How much energy is wasted?
c) What is the efficiency of the catapult?

28) What are the standard units of electrical energy?
29)*Calculate how many kWh of electrical energy are used by a 0.5 kW heater used for 15 minutes.
30) Would a battery-powered radio or a clockwork radio be more suitable to use in rural Africa? Why?

* Answers on p.108.

Energy Sources & Power Stations

There are 12 different types of energy resource.
They fit into two broad types: renewable and non-renewable.

Non-Renewable Energy Resources Will Run Out One Day

The non-renewables are the three FOSSIL FUELS and NUCLEAR:

1) Coal
2) Oil
3) Natural gas
4) Nuclear fuels (uranium and plutonium)

a) They will all 'run out' one day.
b) They all do damage to the environment.
c) But they provide most of our energy.

Renewable Energy Resources Will Never Run Out

The renewables are:

1) Wind
2) Waves
3) Tides
4) Hydroelectric
5) Solar
6) Geothermal
7) Food
8) Biofuels

a) These will never run out.
b) Most of them do damage the environment, but in less nasty ways than non-renewables.
c) The trouble is they don't provide much energy and some of them are unreliable because they depend on the weather.

Energy Sources can be Burned to Drive Turbines in Power Stations

Most of the electricity we use is generated from the four NON-RENEWABLE sources of energy (coal, oil, gas and nuclear) in big power stations, which are all pretty much the same apart from the boiler.

Learn the basic features of the typical power station shown here and also the nuclear reactor below.

1) The fossil fuel is burned to convert its stored chemical energy into heat (thermal) energy.
2) The heat energy is used to heat water (or air in some fossil-fuel power stations) to produce steam.

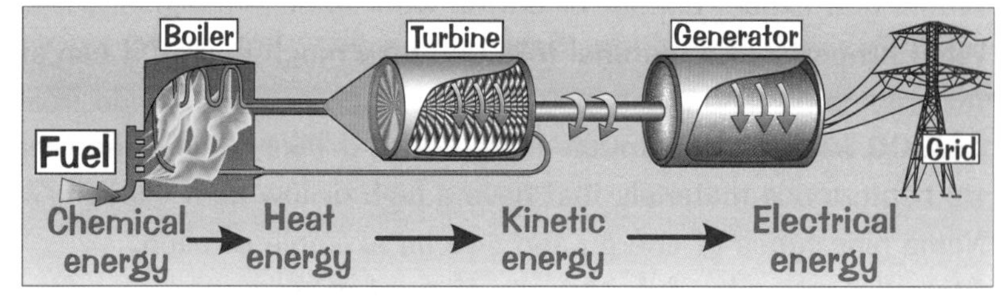

3) The steam turns a turbine, converting heat energy into kinetic energy.
4) The turbine is connected to a generator, which transfers kinetic energy into electrical energy.

Nuclear Reactors are Just Fancy Boilers

1) A nuclear power station is mostly the same as the one above, but with nuclear fission of uranium or plutonium producing the heat to make steam to drive turbines, etc. The difference is in the boiler, as shown here:
2) Nuclear power stations take the longest time of all the power stations to start up. Natural gas power stations take the shortest time of all the fossil fuel power stations.

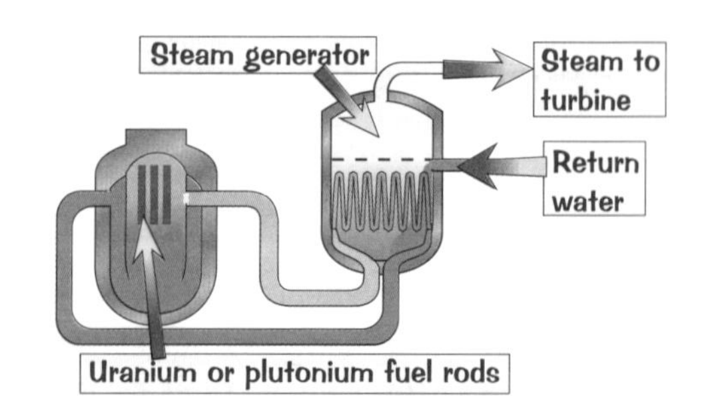

It all boils down to steam...

Steam engines were invented as long ago as the 17th century, and yet we're still using that idea to produce most of our electricity today, over 300 years later. Amazing...

Renewable Energy Sources

Renewable energy sources, like wind, waves and solar energy, will not run out. What's more, they do a lot less damage to the environment. They don't generate as much electricity as non-renewables though — if they did we'd all be using solar-powered toasters by now.

Wind Power — Lots of Little Wind Turbines

1) This involves putting lots of windmills (wind turbines) up in exposed places like on moors or round coasts.
2) Each wind turbine has its own generator inside it. The electricity is generated directly from the wind turning the blades, which turn the generator.

3) There's no pollution (except for a little bit when they're manufactured).
4) But they do spoil the view. You need about 1500 wind turbines to replace one coal-fired power station and 1500 of them cover a lot of ground — which would have a big effect on the scenery.
5) And they can be very noisy, which can be annoying for people living nearby.
6) There's also the problem of no power when the wind stops, and it's impossible to increase supply when there's extra demand.
7) The initial costs are quite high, but there are no fuel costs and minimal running costs.
8) There's no permanent damage to the landscape — if you remove the turbines, you remove the noise and the view returns to normal.

Solar Cells — Expensive but No Environmental Damage

(well, there may be a bit caused by making the cells)

1) Solar cells generate electric currents directly from sunlight. Solar cells are often the best source of energy for calculators and watches which don't use much electricity.

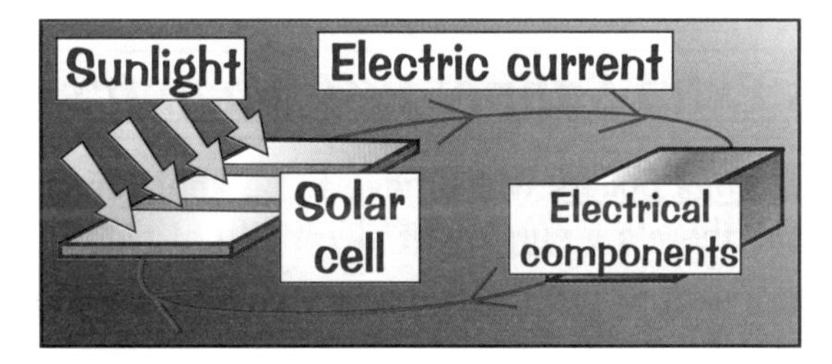

2) Solar power is often used in remote places where there's not much choice (e.g. the Australian outback) and to power electric road signs and satellites.
3) There's no pollution. (Although they do use quite a lot of energy to manufacture in the first place.)
4) In sunny countries solar power is a very reliable source of energy — but only in the daytime. Solar power can still be cost-effective even in cloudy countries like Britain.
5) Initial costs are high but after that the energy is free and running costs almost nil.
6) Solar cells are usually used to generate electricity on a relatively small scale, e.g. powering individual homes.
7) It's often not practical or too expensive to connect them to the National Grid — the cost of connecting them to the National Grid can be enormous compared with the value of the electricity generated.

People love the idea of wind power — just not in their back yard...

Did you know you can now get rucksacks with built-in solar cells to charge up your mobile phone, MP3 player and digital camera while you're wandering around. Pretty cool, huh.

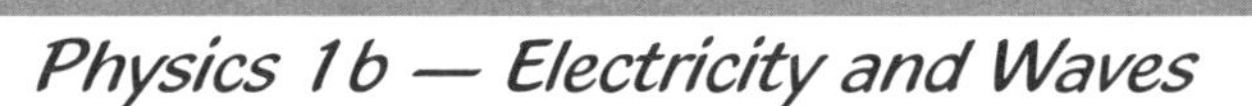

Renewable Energy Sources

Good ol' water. Not only can we drink it — we can also use it to turn turbines in the same way as wind. Wherever water is moving — in waves, rivers and tides, we can transfer its kinetic energy into electrical energy.

Hydroelectric Power Uses Falling Water

1) Hydroelectric power usually requires the flooding of a valley by building a big dam.
2) Rainwater is caught and allowed out through turbines. There is no pollution (as such).
3) But there is a big impact on the environment due to the flooding of the valley (rotting vegetation releases methane and CO_2) and possible loss of habitat for some species (sometimes the loss of whole villages). The reservoirs can also look very unsightly when they dry up.
   Putting hydroelectric power stations in remote valleys tends to reduce their impact on humans.

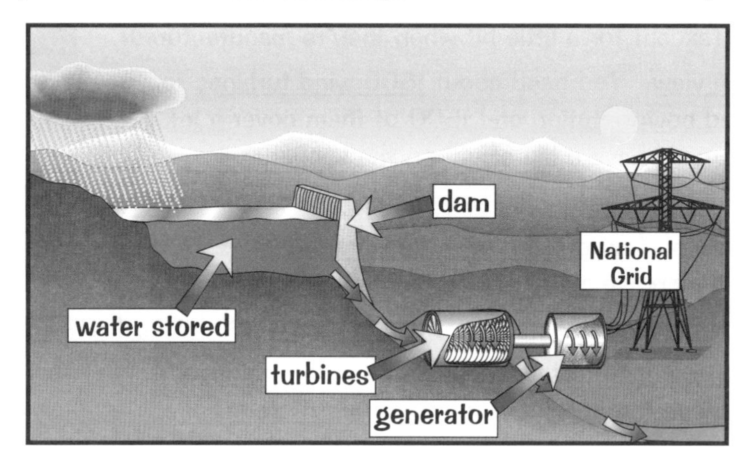

4) A big advantage is it can provide an immediate response to an increased demand for electricity.
5) There's no problem with reliability except in times of drought — but remember this is **Great Britain** we're talking about.
6) Initial costs are high, but there's no fuel and minimal running costs.
7) It can be a useful way to generate electricity on a small scale in remote areas.

Pumped Storage Gives Extra Supply Just When It's Needed

1) Most large power stations have huge boilers which have to be kept running all night even though demand is very low. This means there's a surplus of electricity at night.
2) It's surprisingly difficult to find a way of storing this spare energy for later use.
3) Pumped storage is one of the best solutions.
4) In pumped storage, 'spare' night-time electricity is used to pump water up to a higher reservoir.
5) This can then be released quickly during periods of peak demand such as at teatime each evening, to supplement the steady delivery from the big power stations.
6) Remember, pumped storage uses the same idea as hydroelectric power, but it isn't a way of generating power — it's simply a way of storing energy which has already been generated.

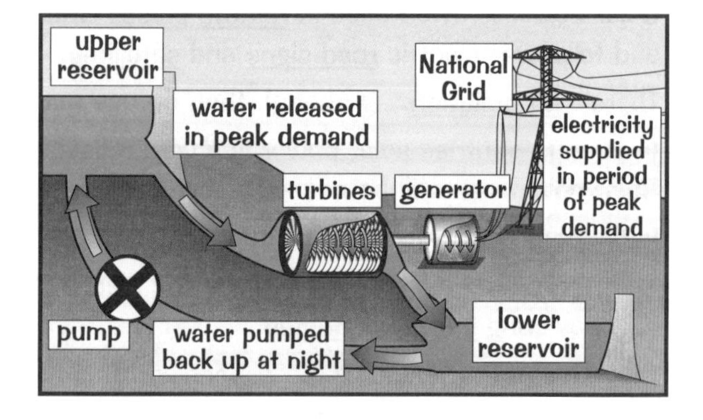

The hydroelectric power you're supplying — it's electrifying...

In Britain only a pretty small percentage of our electricity comes from hydroelectric power at the moment, but in some other parts of the world they rely much more heavily on it. For example, in the last few years, 99% of Norway's energy came from hydroelectric power. 99% — that's huge!

Renewable Energy Sources

Don't worry — I haven't forgotten about wave power and tidal power. It's easy to get confused between these two just because they're both to do with the seaside — but don't. They are completely different.

Wave Power — Lots of Little Wave-Powered Turbines

1) You need lots of small wave-powered turbines located around the coast.
2) As waves come in to the shore they provide an up and down motion which can be used to drive a generator.
3) There is no pollution. The main problems are spoiling the view and being a hazard to boats.
4) They are fairly unreliable, since waves tend to die out when the wind drops.
5) Initial costs are high, but there are no fuel costs and minimal running costs. Wave power is never likely to provide energy on a large scale, but it can be very useful on small islands.

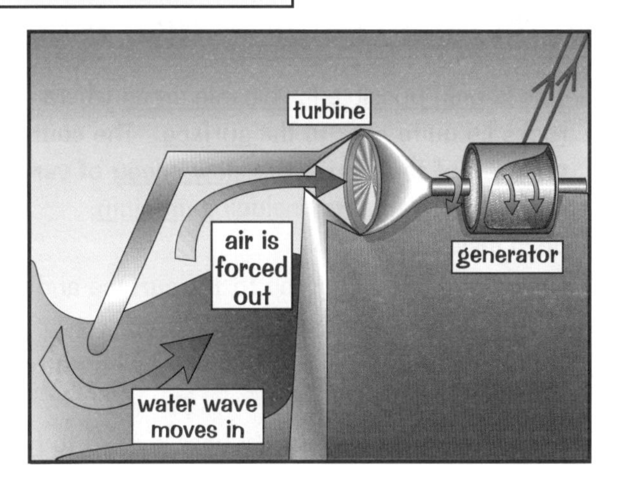

Tidal Barrages — Using the Sun and Moon's Gravity

1) Tidal barrages are big dams built across river estuaries, with turbines in them.
2) As the tide comes in it fills up the estuary to a height of several metres — it also drives the turbines. This water can then be allowed out through the turbines at a controlled speed.
3) The source of the energy is the gravity of the Sun and the Moon.
4) There is no pollution. The main problems are preventing free access by boats, spoiling the view and altering the habitat of the wildlife, e.g. wading birds, sea creatures and beasties who live in the sand.

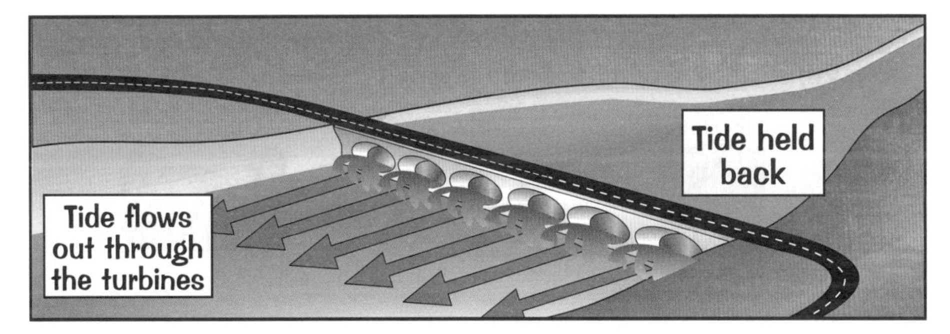

5) Tides are pretty reliable in the sense that they happen twice a day without fail, and always near to the predicted height. The only drawback is that the height of the tide is variable so lower (neap) tides will provide significantly less energy than the bigger 'spring' tides. They also don't work when the water level is the same either side of the barrage — this happens four times a day because of the tides. But tidal barrages are excellent for storing energy ready for periods of peak demand.
6) Initial costs are moderately high, but there are no fuel costs and minimal running costs. Even though it can only be used in some of the most suitable estuaries tidal power has the potential for generating a significant amount of energy.

Learn about Wave Power — and bid your cares goodbye...

I do hope you appreciate the big big differences between tidal power and wave power. They both involve salty seawater, sure — but there the similarities end. Lots of jolly details then, just waiting to be absorbed into your cavernous intracranial void. Smile and enjoy. And learn.

Renewable Energy Sources

Well, who'd know it — there's yet more energy lurking about in piles of rubbish and deep underground. Makes you wonder sometimes why we even need to use oil. (If you are wondering about that, page 32 is all about comparing energy resources, so sit tight for now.)

Geothermal Energy — Heat from Underground

1) This is only possible in volcanic areas where hot rocks lie quite near to the surface. The source of much of the heat is the slow decay of various radioactive elements, including uranium, deep inside the Earth.
2) Steam and hot water rise to the surface and are used to drive a generator.
3) This is actually brilliant free energy with no real environmental problems.
4) In some places, geothermal heat is used to heat buildings directly, without being converted to electrical energy.
5) The main drawback with geothermal energy is there aren't very many suitable locations for power plants.
6) Also, the cost of building a power plant is often high compared to the amount of energy we can get out of it.

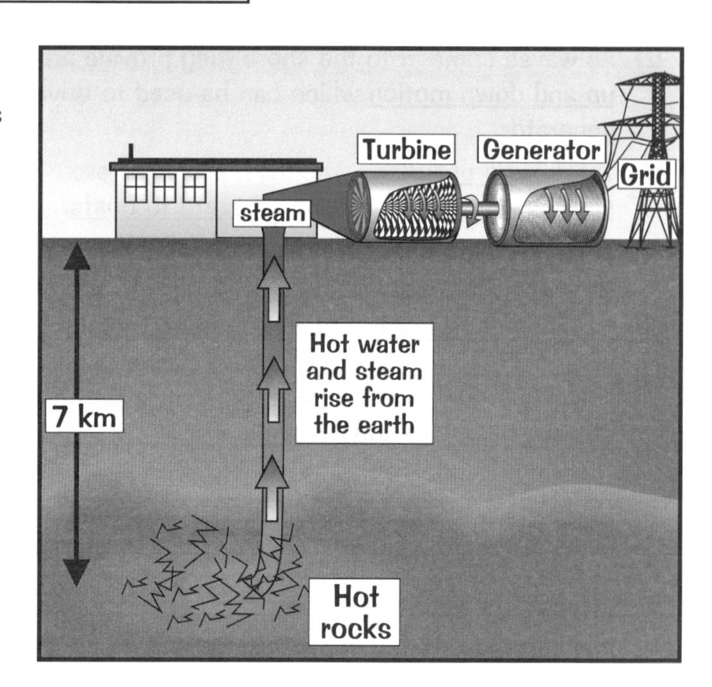

Biofuels are Made from Plants and Waste

1) Biofuels are renewable energy resources. They're used to generate electricity in exactly the same way as fossil fuels (see p.26) — they're burnt to heat up water.

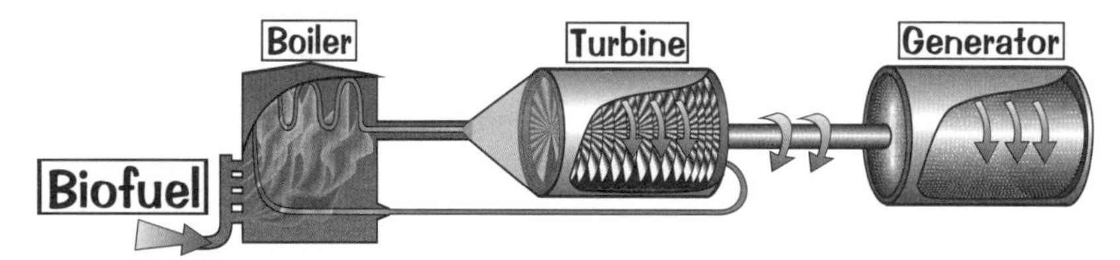

2) They can be also used in some cars — just like fossil fuels.
3) Biofuels can be solids (e.g. straw, nutshells and woodchips), liquids (e.g. ethanol) or gases (e.g. methane 'biogas' from sludge digesters).
4) We can get biofuels from organisms that are still alive or from dead organic matter — like fossil fuels, but from organisms that have been living much more recently.
5) E.g. crops like sugar cane can be fermented to produce ethanol, or plant oils can be modified to produce biodiesel.

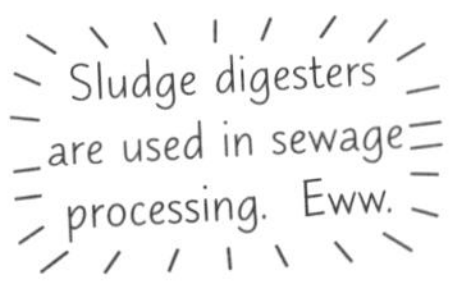

Sugar cane to ethanol — a terrible waste in my opinion...

Biofuels sound quite futuristic. But believe it or not, biofuel mixed with petrol or diesel was actually used in some cars before WW2. Biofuel never really became massively successful though because of cheap oil. One big advantage of biofuels is they don't release as much greenhouse gas compared with common transport fuels like petrol and diesel. They aren't completely innocent in the pollution game though, as you'll see on the next page.

Energy Sources and the Environment

They might fly you to Spain for your holidays and power your games consoles, but using non-renewable energy sources and biofuels to generate electricity can have damaging effects on the environment.

Non-Renewables are Also Linked to Other Environmental Problems

1) All three fossil fuels (coal, oil and gas) release CO_2 into the atmosphere when they're burned. For the same amount of energy produced, coal releases the most CO_2, followed by oil then gas. All this CO_2 adds to the greenhouse effect, and contributes to global warming.
2) Burning coal and oil releases sulfur dioxide, which causes acid rain. Acid rain can be harmful to trees and soils and can have far-reaching effects in ecosystems.
3) Acid rain can be reduced by taking the sulfur out before the fuel is burned, or cleaning up the emissions.
4) Coal mining makes a mess of the landscape, especially "open-cast mining".
5) Oil spillages cause serious environmental problems, affecting mammals and birds that live in and around the sea. We try to avoid them, but they'll always happen.
6) Nuclear power is clean but the nuclear waste is very dangerous and difficult to dispose of.
7) Nuclear fuel (i.e. uranium) is relatively cheap but the overall cost of nuclear power is high due to the cost of the power plant and final decommissioning.
8) Nuclear power always carries the risk of a major catastrophe like the Chernobyl disaster in 1986.

Biofuels Have Their Disadvantages Too

1) Biofuels (see p.30) are a relatively quick and 'natural' source of energy and are supposedly carbon neutral.
2) There is still debate into the impact of biofuels on the environment, once the full energy that goes into the production is considered.

> The plants that grew to produce the waste (or to feed the animals that produced the dung) absorbed carbon dioxide from the atmosphere as they were growing. When the waste is burnt, this CO_2 is re-released into the atmosphere. So it has a neutral effect on atmospheric CO_2 levels (although this only really works if you keep growing plants at the same rate you're burning things). Biofuel production also creates methane emissions — a lot of this comes from the animals. Nice.

Huge areas of land are needed to produce biofuels on a large scale.

3) In some regions, large areas of forest have been cleared to make room to grow biofuels, resulting in lots of species losing their natural habitats. The decay and burning of this vegetation also increases CO_2 and methane emissions.
4) Biofuels have potential, but their use is limited by the amount of available farmland that can be dedicated to their production.

Carbon Capture can Reduce the Impact of Carbon Dioxide

1) Carbon capture and storage (CCS) is used to reduce the amount of CO_2 building up in the atmosphere and reduce the strength of the greenhouse effect.
2) CCS works by collecting the CO_2 from power stations before it is released into the atmosphere.
3) The captured CO_2 can then be pumped into empty gas fields and oil fields like those under the North Sea. It can be safely stored without it adding to the greenhouse effect.
4) CCS is a new technology that's developing quickly. New ways of storing CO_2 are being explored, including storing CO_2 dissolved in seawater at the bottom of the ocean and capturing CO_2 with algae, which can then be used to produce oil that can be used as a biofuel.

Biofuels are great — but don't burn your biology notes just yet...

Wowsers. There certainly is a lot to bear in mind with all the different energy sources and all the good things and nasty things associated with each of them. The next page is really handy for making comparisons between different energy sources — it'll tell you everything you need to know. (Secret hint: you should definitely read it.)

Comparison of Energy Resources

Setting Up a Power Station

Because coal and oil are running out fast, many old coal- and oil-fired power stations are being taken out of use. Often they're being replaced by gas-fired power stations because they're quick to set up, there's still quite a lot of gas left and gas doesn't pollute as badly as coal and oil.

But gas is not the only option, as you really ought to know if you've been concentrating at all over the last few pages.

When looking at the options for a new power station, there are several factors to consider: How much it costs to set up and run, how long it takes to build, how much power it can generate, etc. Then there are also the trickier factors like damage to the environment and impact on local communities. And because these are often very contentious issues, getting permission to build certain types of power station can be a long-running process, and hence increase the overall set-up time. The time and cost of decommissioning (shutting down) a power plant can also be a crucial factor.

Set-Up Costs

Renewable resources often need bigger power stations than non-renewables for the same output. And as you'd expect, the bigger the power station, the more expensive.

Nuclear reactors and hydroelectric dams also need huge amounts of engineering to make them safe, which bumps up the cost.

Reliability Issues

All the non-renewables are reliable energy providers (until they run out).

Many of the renewable sources depend on the weather, which means they're pretty unreliable here in the UK. The exceptions are tidal power and geothermal (which don't depend on weather).

Environmental Issues

If there's a fuel involved, there'll be waste pollution and you'll be using up resources.

If it relies on the weather, it's often got to be in an exposed place where it sticks out like a sore thumb.

Atmospheric Pollution
Coal, Oil, Gas,
(+ others, though less so)

Visual Pollution
Coal, Oil, Gas, Nuclear, Tidal, Waves, Wind, Hydroelectric,

Other Problems
Nuclear (dangerous waste, explosions, contamination), Hydroelectric (dams bursting)

Using Up Resources
Coal, Oil, Gas, Nuclear

Noise Pollution
Coal, Oil, Gas, Nuclear, Wind,

Disruption of Habitats
Hydroelectric, Tidal, Biofuels.

Disruption of Leisure Activities (e.g. boats)
Waves, Tidal

Set-Up/Decommissioning Time

These are both affected by the size of the power station, the complexity of the engineering and also the planning issues (e.g. discussions over whether a nuclear power station should be built on a stretch of beautiful coastline can last years). Gas is one of the quickest to set up. Nuclear power stations take by far the longest (and cost the most) to decommission.

Running/Fuel Costs

Renewables usually have the lowest running costs, because there's no actual fuel involved.

Location Issues

This is fairly common sense — a power station has to be near to the stuff it runs on.

Solar — pretty much anywhere, though the sunnier the better

Gas — pretty much anywhere there's piped gas (most of the UK)

Hydroelectric — hilly, rainy places with floodable valleys, e.g. the Lake District, Scottish Highlands

Wind — exposed, windy places like moors and coasts or out at sea

Oil — near the coast (oil transported by sea)

Waves — on the coast

Coal — near coal mines, e.g. Yorkshire, Wales

Nuclear — away from people (in case of disaster), near water (for cooling)

Tidal — big river estuaries where a dam can be built

Geothermal — fairly limited, only in places where hot rocks are near the Earth's surface

Of course — the biggest problem is we need too much electricity...

It would be lovely if we could get rid of all the nasty polluting power stations and replace them with clean, green fuel, just like that... but it's not quite that simple. Renewable energy has its own problems too, and probably isn't enough to power the whole country without having a wind farm in everyone's back yard.

Electricity and the National Grid

The National Grid is the network of pylons and cables that covers the whole of Britain, getting electricity to homes everywhere. Whoever you pay for your electricity, it's the National Grid that gets it to you.

Electricity is Distributed via the National Grid...

1) The National Grid takes electrical energy from power stations to where it's needed in homes and industry.
2) It enables power to be generated anywhere on the grid, and then be supplied anywhere else on the grid.
3) To transmit the huge amount of power needed, you need either a high voltage or a high current.
4) The problem with a high current is that you lose loads of energy through heat in the cables.
5) It's much cheaper to boost the voltage up really high (to 400 000 V) and keep the current very low.

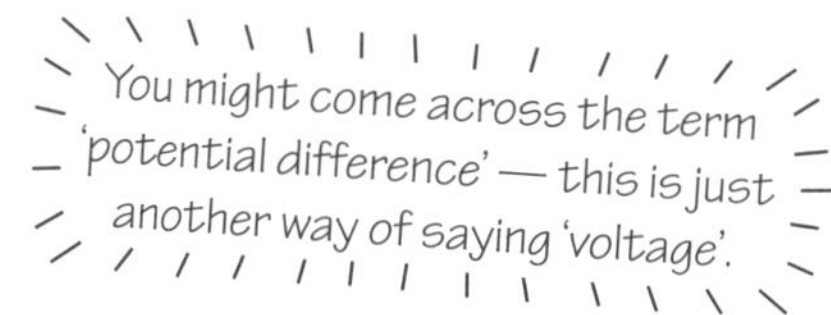

...With a Little Help from Pylons and Transformers

1) To get the voltage to 400 000 V to transmit power requires transformers as well as big pylons with huge insulators — but it's still cheaper.
2) The transformers have to step the voltage up at one end, for efficient transmission, and then bring it back down to safe, usable levels at the other end.
3) The voltage is increased ('stepped up') using a step-up transformer. (Yep, does what it says on the tin.)
4) It's then reduced again ('stepped down') at the consumer end using a step-down transformer.

There are Different Ways to Transmit Electricity

1) Electrical energy can be moved around by cables buried in the ground, as well as in overhead power lines.
2) Each of these different options has its pros and cons:

	Setup cost	Maintenance	Faults	How it looks	Affected by weather	Reliability	How easy to set up	Disturbance to land
Overhead Cables	lower	lots needed	easy to access	ugly	yes	less reliable	easy	minimal
Underground Cables	higher	minimal	hard to access	hidden	no	more reliable	hard	lots

Supply and Demand

1) The National Grid needs to generate and direct all the energy that the country needs — our energy demands keep on increasing too.
2) In order to meet these demands in the future, the energy supplied to the National Grid will need to increase, or the energy demands of consumers will need to decrease.
3) In the future, supply can be increased by opening more power plants or increasing their power output (or by doing both).
4) Demand can be reduced by consumers using more energy-efficient appliances, and being more careful not to waste energy in the home (e.g. turning off the lights or running washing machines at cooler temperatures).

Transformers — NOT robots in disguise...

You don't need to know the details about exactly what transformers are and how they work — just that they increase and decrease the voltage to minimise power losses in the National Grid. Make sure you know the good, bad and occasionally ugly pros and cons of underground and over-ground electricity transmission too.

Wave Basics

Waves transfer energy from one place to another without transferring any matter (stuff).

Waves Have Amplitude, Wavelength and Frequency

1) The amplitude is the displacement from the rest position to the crest (NOT from a trough to a crest).
2) The wavelength is the length of a full cycle of the wave, e.g. from crest to crest.
3) Frequency is the number of complete waves passing a certain point per second OR the number of waves produced by a source each second. Frequency is measured in hertz (Hz). 1 Hz is 1 wave per second.

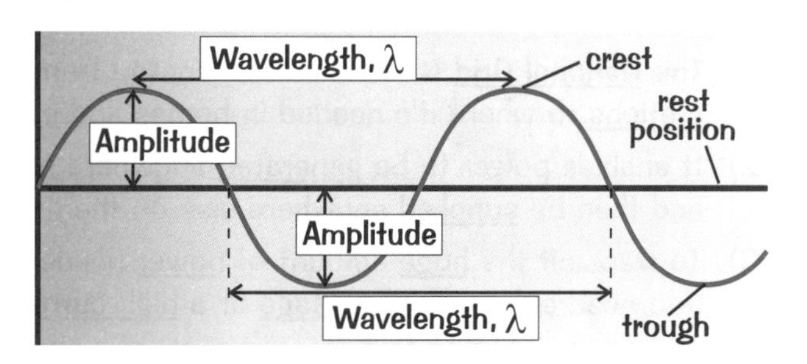

Transverse Waves Have Sideways Vibrations

Most waves are transverse:

1) Light and all other EM waves.
2) Ripples on water.
3) Waves on strings.
4) A slinky spring wiggled up and down.

In TRANSVERSE waves the vibrations are PERPENDICULAR (at 90°) to the DIRECTION OF ENERGY TRANSFER of the wave.

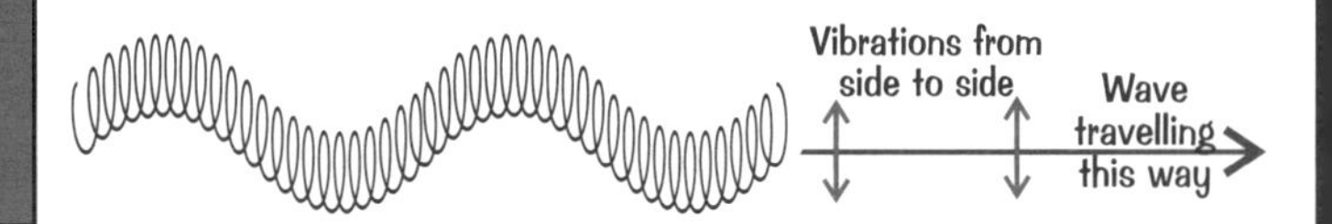

Longitudinal Waves Have Vibrations Along the Same Line

Examples of longitudinal waves are:

1) Sound waves and ultrasound.
2) Shock waves, e.g. seismic waves.
3) A slinky spring when you push the end.

Water waves, shock waves and waves in springs and ropes are all examples of mechanical waves.

In LONGITUDINAL waves the vibrations are PARALLEL to the DIRECTION OF ENERGY TRANSFER of the wave.

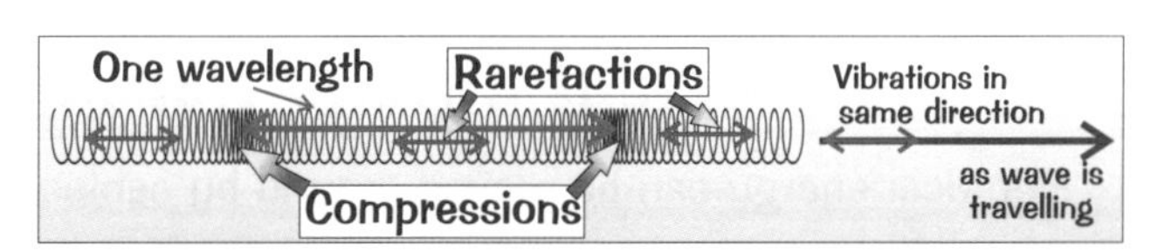

Wave Speed = Frequency × Wavelength

The equation below applies to all waves. You need to learn it — and practise using it.

Speed = Frequency × Wavelength
(m/s) (Hz) (m)

OR $v = f \times \lambda$

Speed (v is for velocity)
Frequency
Wavelength (that's the Greek letter 'lambda')

EXAMPLE: A radio wave has a frequency of 92.2×10^6 Hz. Find its wavelength. (The speed of all EM waves is 3×10^8 m/s.)

ANSWER: You're trying to find λ using f and v, so you've got to rearrange the equation. So $\lambda = v \div f = 3 \times 10^8 \div 9.22 \times 10^7 = 3.25$ m.

The speed of a wave is usually independent of the frequency or amplitude of the wave.

Waves — dig the vibes, man...

The first thing to learn is that diagram at the top of the page. Then get that $v = f \times \lambda$ business imprinted on your brain. When you've done that, try this question: A sound wave travelling in a solid has a frequency of 1.9×10^4 Hz and a wavelength of 12.5 cm. Find its speed.*

* Answers on p.108.

Waves Properties

If you're anything like me, you'll have spent hours gazing into a mirror in wonder. Here's why...

All Waves Can be Reflected, Refracted and Diffracted

1) When waves arrive at an obstacle (or meet a new material), their direction of travel can be changed.
2) This can happen by reflection (see below) or by refraction or diffraction (see page 36).

Reflection of Light Lets Us See Things

1) Reflection of light is what allows us to see objects. Light bounces off them into our eyes.
2) When light travelling in the same direction reflects from an uneven surface such as a piece of paper, the light reflects off at different angles.
3) When light travelling in the same direction reflects from an even surface (smooth and shiny like a mirror) then it's all reflected at the same angle and you get a clear reflection.

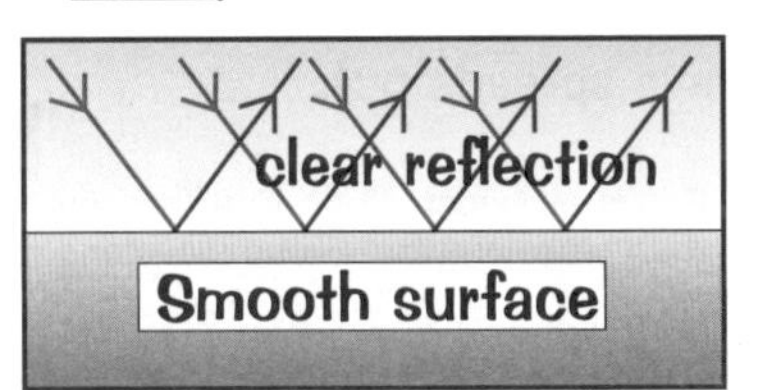

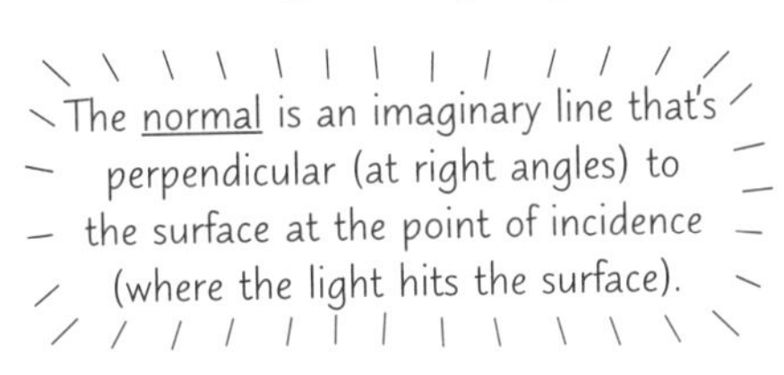

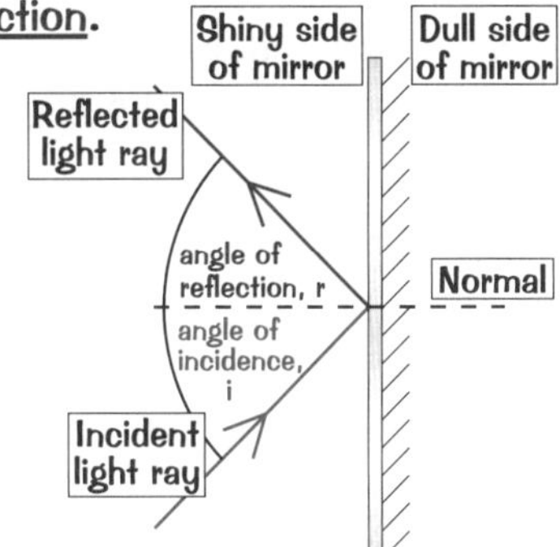

4) The LAW OF REFLECTION applies to every reflected ray:

Angle of INCIDENCE = Angle of REFLECTION

Note that these two angles are ALWAYS defined between the ray itself and the NORMAL, dotted above. Don't ever label them as the angle between the ray and the surface. Definitely uncool.

Draw a Ray Diagram for an Image in a Plane Mirror

You need to be able to reproduce this entire diagram of how an image is formed in a PLANE MIRROR. Learn these important points:

1) The image is the same size as the object.
2) It is AS FAR BEHIND the mirror as the object is in front.
3) The image is virtual and upright. The image is virtual because the object appears to be behind the mirror.
4) The image is laterally inverted — the left and right sides are swapped, i.e. the object's left side becomes its right side in the image.

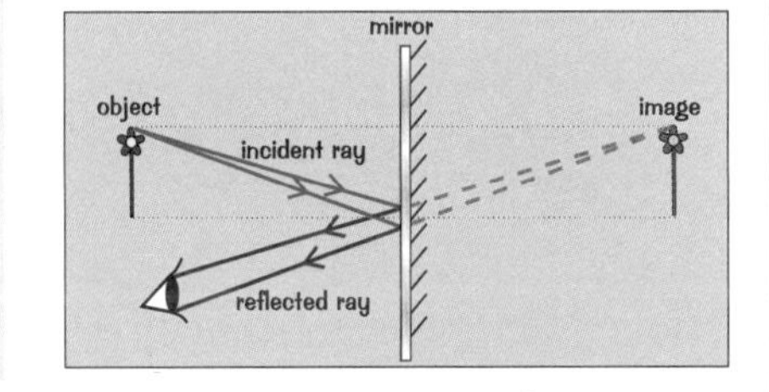

1) First off, draw the virtual image. Don't try to draw the rays first. Follow the rules in the above box — the image is the same size, and it's as far behind the mirror as the object is in front.

2) Next, draw a reflected ray going from the top of the virtual image to the top of the eye. Draw a bold line for the part of the ray between the mirror and eye, and a dotted line for the part of the ray between the mirror and virtual image.

3) Now draw the incident ray going from the top of the object to the mirror. The incident and reflected rays follow the law of reflection — but you don't actually have to measure any angles. Just draw the ray from the object to the point where the reflected ray meets the mirror.

4) Now you have an incident ray and reflected ray for the top of the image. Do steps 2 and 3 again for the bottom of the eye — a reflected ray going from the image to the bottom of the eye, then an incident ray from the object to the mirror.

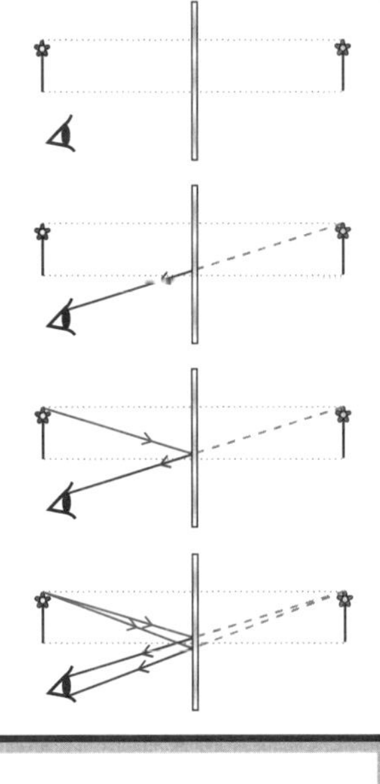

Plane mirrors — what pilots use to look behind them...

Make sure you can draw clear ray diagrams and you'll be well on your way to picking up lotsa marks in the exam.

Refraction and Diffraction

If you thought reflection was good, you'll just love diffraction and refraction — it's awesome. If you didn't find reflection interesting then I'm afraid it's tough luck — you need to know about all three of them. Sorry.

Diffraction and Refraction are a Bit More Complicated

1) Reflection's quite straightforward, but there are other ways that waves can be made to change direction.
2) They can be refracted — which means they go through a new material but change direction.
3) Or they can be diffracted — the waves 'bend round' obstacles, causing the waves to spread out.

Diffraction — Waves Spreading Out

1) All waves spread out ('diffract') at the edges when they pass through a gap or pass an obstacle.
2) The amount of diffraction depends on the size of the gap relative to the wavelength of the wave. The narrower the gap, or the longer the wavelength, the more the wave spreads out.
3) A narrow gap is one that is the same order of magnitude as the wavelength of the wave — i.e. they're about the same size.
4) So whether a gap counts as narrow or not depends on the wave in question.
5) Light has a very small wavelength (about 0.0005 mm), so it can be diffracted but it needs a really small gap.

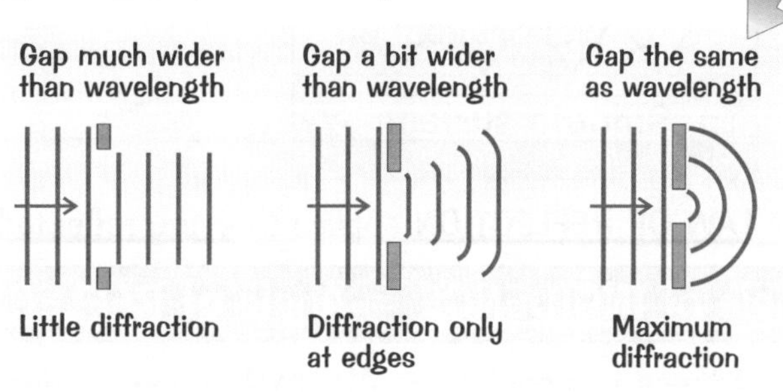

Refraction — Changing the Speed of a Wave Can Change its Direction

1) When a wave crosses a boundary between two substances (from glass to air, say) it changes direction:

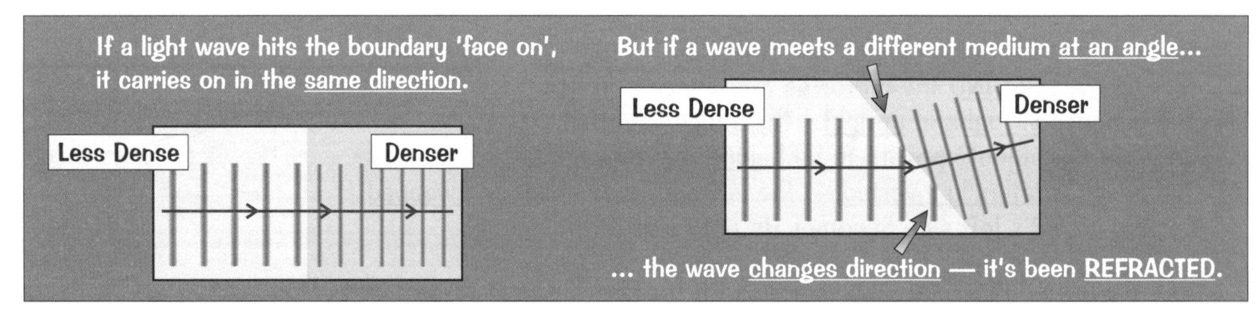

2) When light shines on a glass window pane, some of the light is reflected, but a lot of it passes through the glass and gets refracted as it does so.
3) Waves are only refracted if they meet a new medium at an angle.
4) If they're travelling along the normal (i.e. the angle of incidence is zero) they will change speed, but are NOT refracted — they don't change direction.

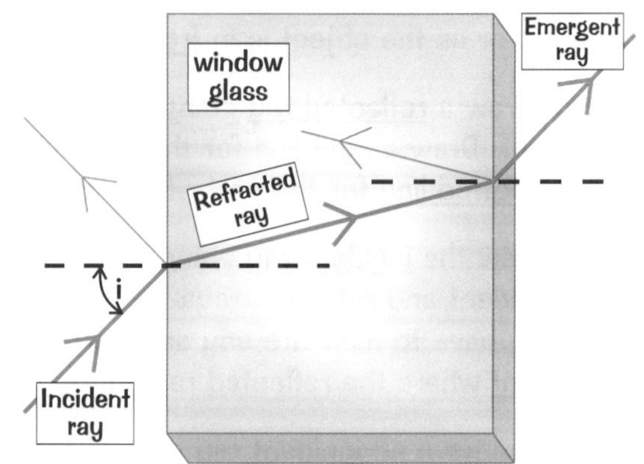

Lights, camera, refraction...

Diffraction's not too hard to get to grips with — especially if you can remember those diagrams at the top of the page. Also remember that all waves can be diffracted — so it doesn't matter if they're longitudinal or transverse waves. The key point to remember about refraction is that the wave has to meet a boundary at an angle.

EM Waves and Communication

Types of electromagnetic (EM) wave have a lot in common with one another, but their differences make them useful to us in different ways. These pages are packed with loads of dead important info, so pay attention...

There's a Continuous Spectrum of EM Waves

EM waves with different wavelengths (or frequencies) have different properties. We group them into seven basic types, but the different regions actually merge to form a continuous spectrum.

They're shown below with increasing frequency and energy (decreasing wavelength) from left to right.

	RADIO WAVES	MICRO WAVES	INFRA RED	VISIBLE LIGHT	ULTRA VIOLET	X-RAYS	GAMMA RAYS
wavelength →	1 m – 10^4 m	10^{-2} m (1 cm)	10^{-5} m (0.01 mm)	10^{-7} m	10^{-8} m	10^{-10} m	10^{-15} m

1) EM waves vary in wavelength from around 10^{-15} m to more than 10^4 m.
2) All the different types of EM wave travel at the same speed (3×10^8 m/s) in a vacuum (e.g. space).
3) EM waves with higher frequencies have shorter wavelengths.
4) Because of their different properties, different EM waves are used for different purposes.

Radio Waves are Used Mainly for Communication

1) Radio waves are EM radiation with wavelengths longer than about 10 cm.
2) Long-wave radio (wavelengths of 1 – 10 km) can be transmitted from London, say, and received halfway round the world. That's because long wavelengths diffract (bend, see p.36) around the curved surface of the Earth.

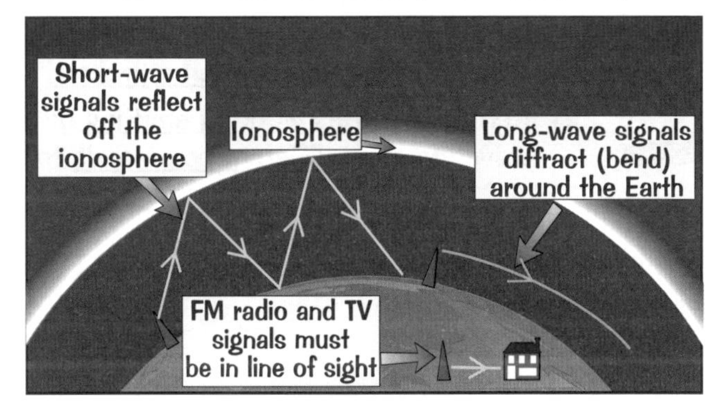

3) Long-wave radio wavelengths can also diffract around hills, into tunnels and all sorts.
4) This diffraction effect makes it possible for radio signals to be received even if the receiver isn't in line of the sight of the transmitter.
5) The radio waves used for TV and FM radio transmissions have very short wavelengths (10 cm – 10 m). To get reception, you must be in direct sight of the transmitter — the signal doesn't bend around hills or travel far through buildings.
6) Short-wave radio signals (wavelengths of about 10 m – 100 m) can, like long-wave, be received at long distances from the transmitter. That's because they are reflected (see p.35) from the ionosphere — an electrically charged layer in the Earth's upper atmosphere.
7) Medium-wave signals (well, the shorter ones) can also reflect from the ionosphere, depending on atmospheric conditions and the time of day.

Size matters — and my wave's longer than yours...

You'll have to be able to name the order of the different types of EM waves in terms of their energy, frequency and wavelength. To remember the order of increasing frequency and energy, I use the mnemonic Rock Music Is Very Useful for eXperiments with Goats. It sounds stupid but it does work — why not make up your own...

EM Waves and Their Uses

Radio waves aren't the only waves used for communication — other EM waves come in pretty handy too. The most important thing is to think about how the properties of a wave relate to its uses.

Microwaves are Used for Satellite Communication and Mobile Phones

1) Communication to and from satellites (including satellite TV signals and satellite phones) uses microwaves. But you need to use microwaves which can pass easily through the Earth's watery atmosphere. Radio waves can't do this.
2) For satellite TV, the signal from a transmitter is transmitted into space...
3) ... where it's picked up by the satellite's receiver dish orbiting thousands of kilometres above the Earth. The satellite transmits the signal back to Earth in a different direction...
4) ... where it's received by a satellite dish on the ground.
5) Mobile phone calls also travel as microwaves between your phone and the nearest transmitter. Some wavelengths of microwaves are absorbed by water molecules and heat them up. If the water in question happens to be in your cells, you might start to cook — so some people think using your mobile a lot (especially next to your head), or living near a mast, could damage your health. There isn't any conclusive evidence either way yet.
6) And microwaves are used by remote-sensing satellites — to 'see' through the clouds and monitor oil spills, track the movement of icebergs, see how much rainforest has been chopped down and so on.

Infrared Waves are Used for Remote Controls and Optical Fibres

1) Infrared waves (see p.37) are used in lots of wireless remote controllers.
2) Remote controls work by emitting different patterns of infrared waves to send different commands to an appliance, e.g. a TV.
3) Optical fibres (e.g. those used in phone lines) can carry data over long distances very quickly.
4) They use both infrared waves and visible light.
5) The signal is carried as pulses of light or infrared radiation and is reflected off the sides of a very narrow core from one end of the fibre to the other.

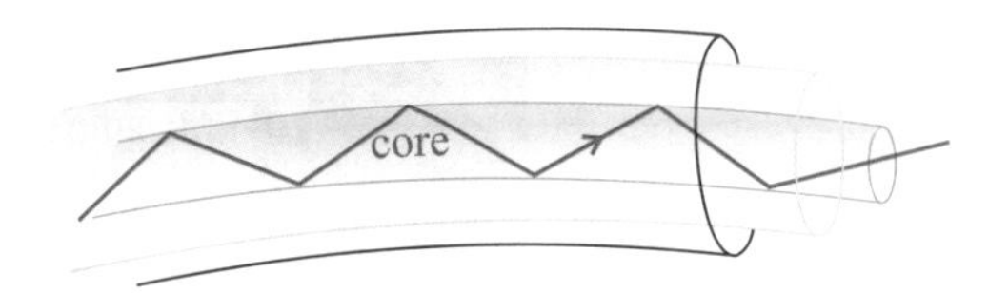

Visible Light is Useful for Photography

It sounds pretty obvious, but photography would be kinda tricky without visible light.

1) Cameras use a lens to focus visible light onto a light-sensitive film or electronic sensor.
2) The lens aperture controls how much light enters the camera (like the pupil in an eye).
3) The shutter speed determines how long the film or sensor is exposed to the light.
4) By varying the aperture and shutter speed (and also the sensitivity of the film or the sensor), a photographer can capture as much or as little light as they want in their photograph.

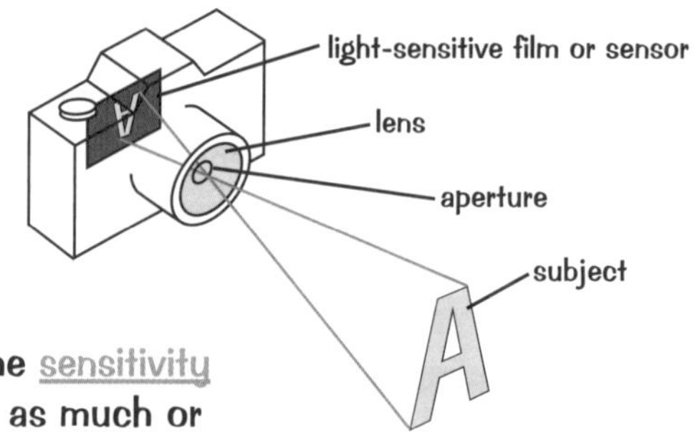

Microwaves are also used for making popcorn — mmm...

I bet you didn't realise that all those different types of technology — like microwaves and infrared — use waves that travel at exactly the same speed as each other (in a vacuum). It's pretty cool stuff.

Sound Waves

We hear sounds when vibrations reach our eardrums. You'll need to know how sound waves work.

Sound Travels as a Wave

1) Sound waves are caused by vibrating objects. These mechanical vibrations are passed through the surrounding medium as a series of compressions. They're a type of longitudinal wave (see page 34).

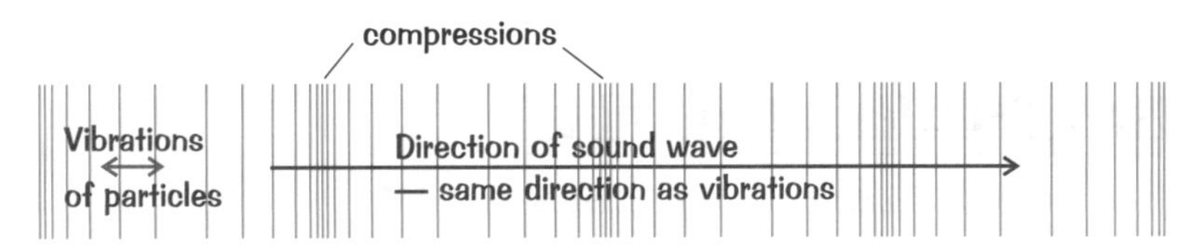

2) Sometimes the sound will eventually travel through someone's inner ear and reach their eardrum, at which point the person might hear it.
3) Because sound waves are caused by vibrating particles, the denser the medium, the faster sound travels through it, generally speaking anyway. Sound generally travels faster in solids than in liquids, and faster in liquids than in gases.
4) Sound can't travel in space, because it's mostly a vacuum (there are no particles).

Sound Waves Can Reflect and Refract

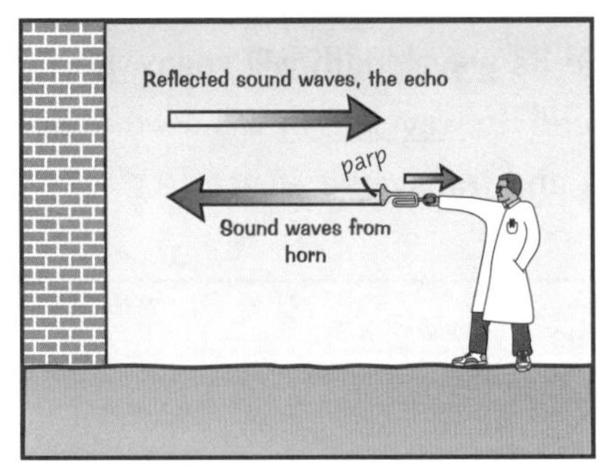

1) Sound waves will be reflected by hard flat surfaces.
2) This is very noticeable in an empty room. A big empty room sounds completely different once you've put carpet, curtains and a bit of furniture in it. That's because these things absorb the sound quickly and stop it echoing around the room. Echoes are just reflected sound waves.
3) You hear a delay between the original sound and the echo because the echoed sound waves have to travel further, and so take longer to reach your ears.

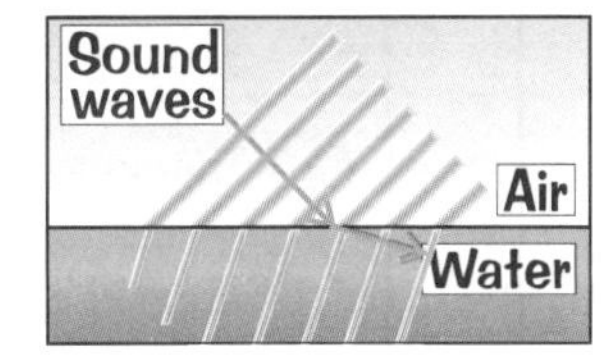

4) Sound waves will also refract (change direction) as they enter different media. As they enter denser material, they speed up. (However, since sound waves are always spreading out so much, the change in direction is hard to spot under normal circumstances.)

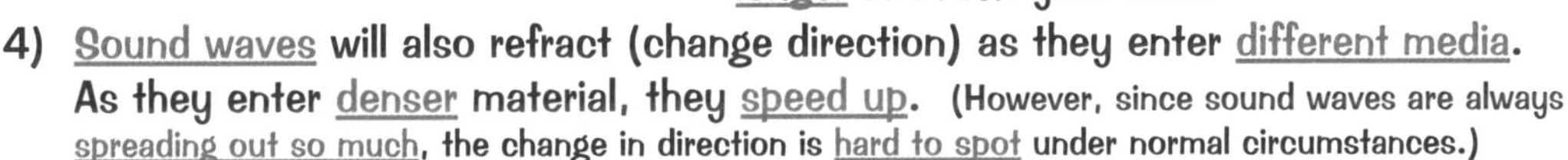

The Higher the Frequency, the Higher the Pitch

1) High frequency sound waves sound high pitched like a squeaking mouse.
2) Low frequency sound waves sound low pitched like a mooing cow.
3) Frequency is the number of complete vibrations each second — so a wave that has a frequency of 100 Hz vibrates 100 times each second.
4) Common units are kHz (1000 Hz) and MHz (1 000 000 Hz).
5) High frequency (or high pitch) also means shorter wavelength (see p.34).
6) The loudness of a sound depends on the amplitude (p.34) of the sound wave. The bigger the amplitude, the louder the sound.

The room always feels big and empty whenever I tell a joke... (It must be the carpets.)

The thing to do here is learn the facts. There's a simple equation that says the more you learn now, the more marks you'll get in the exam. A lot of questions just test whether you've learnt the facts. Easy marks, really. Now journey with me to a distant galaxy to explore the questions that have plagued mankind for centuries...

The Origin of the Universe

OK. Let's not kid ourselves — this is a pretty daunting topic. How the universe started is obviously open to debate, but physicists have got some neat ideas based on their observations of the stars. How romantic...

The Universe Seems to be Expanding

As big as the universe already is, it looks like it's getting even bigger.
All its galaxies seem to be moving away from each other. There's good evidence for this...

1) Light from Other Galaxies is Red-shifted

1) Different chemical elements absorb different frequencies (see p.34) of light.
2) Each element produces a specific pattern of dark lines at the frequencies that it absorbs in the visible spectrum.
3) When we look at light from distant galaxies we can see the same patterns but at slightly lower frequencies than they should be — they're shifted towards the red end of the spectrum. This is called red-shift.
4) It's the same effect as the vrrroomm from a racing car — the engine sounds lower-pitched when the car's gone past you and is moving away from you. This is called the Doppler effect.

An absorption spectrum showing dark lines measured on Earth.

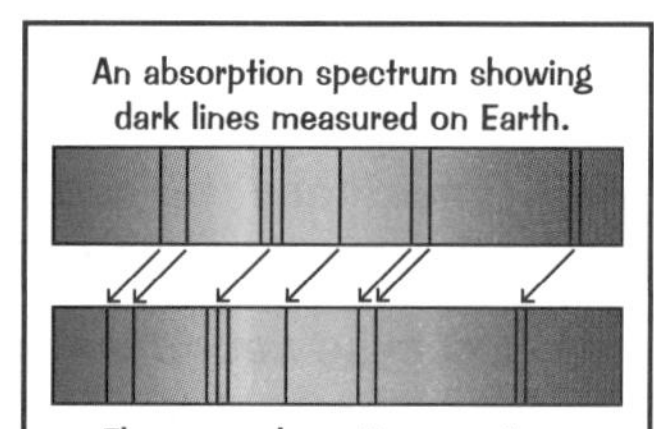

The same absorption spectrum measured from light from a distant galaxy. The dark lines in this spectrum are red-shifted.

The Doppler Effect

1) When something that emits waves moves towards you or away from you, the wavelengths and frequencies of the waves seem different — compared to when the source of the waves is stationary.
2) The frequency of a source moving towards you will seem higher and its wavelength will seem shorter.
3) The frequency of a source moving away from you will seem lower and its wavelength will seem longer.
4) The Doppler effect happens to both longitudinal waves (e.g. sound) and transverse waves (e.g. light and microwaves).

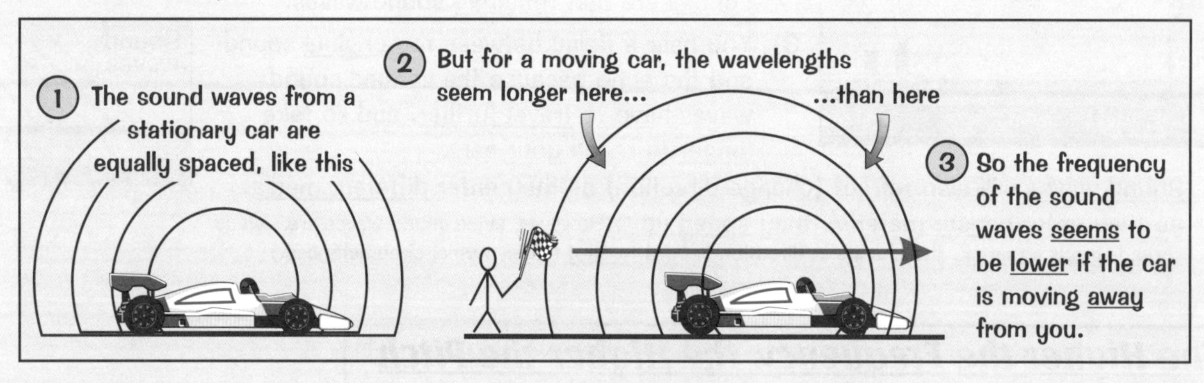

2) The Further Away a Galaxy is, the Greater the Red-shift

1) Measurements of the red-shift suggest that all the galaxies are moving away from us very quickly — and it's the same result whichever direction you look in.
2) More distant galaxies have greater red-shifts than nearer ones.
3) This means that more distant galaxies are moving away from us faster than nearer ones.
4) This provides evidence that the whole universe is expanding.

If a tree falls down in the forest and you're driving away from it...

Listen out for the Doppler effect next time you hear a fast motorbike or a police siren — you should be able to work out if it's coming towards you or speeding away. You can also hear the noise in cartoons when someone falls off a cliff and it plays that classic whistling noise that gets lower, showing them accelerating away from you.

The Origin of the Universe

Once upon a time there was a really Big Bang — that's the most convincing theory we've got.

It All Started Off with a Very Big Bang (Probably)

Right now, distant galaxies are moving away from us — the further away a galaxy is from the us, the faster they're moving away. But something must have got them going. That 'something' was probably a big explosion — so they called it the Big Bang...

1) According to this theory, all the matter and energy in the universe must have been compressed into a very small space. Then it exploded from that single 'point' and started expanding.
2) The expansion is still going on. We can use the current rate of expansion of the universe to estimate its age. Our best guess is that the Big Bang happened about 14 billion years ago.
3) The Big Bang isn't the only game in town. The 'Steady State' theory says that the universe has always existed as it is now, and it always will do. It's based on the idea that the universe appears pretty much the same everywhere. This theory explains the apparent expansion by suggesting that matter is being created in the spaces as the universe expands. But there are some big problems with this theory.
4) The discovery of the cosmic microwave background radiation (CMBR) some years later was strong evidence that the Big Bang was the more likely explanation of the two.

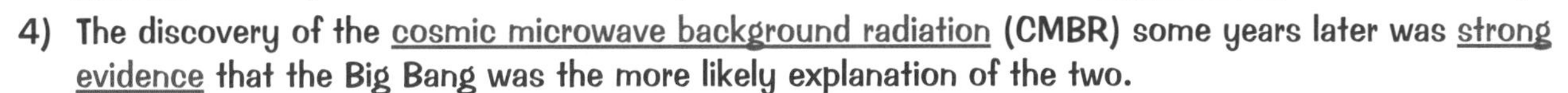

There's a Uniform Microwave Radiation from All Directions

1) Scientists have detected low frequency electromagnetic radiation coming from all parts of the universe.
2) This radiation is largely in the microwave part of the EM spectrum (see p.37). It's known as the cosmic microwave background radiation (CMBR).
3) The Big Bang theory is the only theory that can explain the CMBR.
4) Just after the Big Bang while the universe was still extremely hot, everything in the universe emitted very high frequency radiation. As the universe expanded it has cooled, and this radiation has dropped in frequency and is now seen as microwave radiation.

The Big Bang Theory Has Its Limitations

1) Today nearly all astronomers agree there was a Big Bang. However, there are some who still believe in the Steady State theory. Some of these say the evidence just points that way. Others maybe don't want to change their mind — that would mean admitting they were wrong in the first place.
2) The Big Bang theory isn't perfect. As it stands, it's not the whole explanation of the universe — there are observations that the theory can't yet explain. E.g. for complicated reasons that you don't need to know, the Big Bang theory predicts that the universe's expansion should be slowing down — but as far as we can tell it's actually speeding up.
3) The Big Bang explains the universe's expansion well, but it isn't an explanation for what actually caused the explosion in the first place, or what the conditions were like before the explosion (or if there was a 'before').
4) It seems most likely the Big Bang theory will be adapted in some way to account for its weaknesses rather than just dumped — it explains so much so well that scientists will need a lot of persuading to drop it altogether.

Time and space — it's funny old stuff isn't it...

Proving a scientific theory is impossible. If enough evidence points a certain way, then a theory can look pretty convincing. But that doesn't prove it's a fact — new evidence may change people's minds.

Revision Summary for Physics 1b

It's business time — another chance for you to see which bits went in and which bits you need to flick back and have another read over. You know the drill by now. Do as many of the questions as you can and then try the tricky ones after you've had another chance to read the pages you struggled on. You know it makes sense.

1) What is meant by a non-renewable energy resource?
   Name four different non-renewable energy resources.
2) Explain how electricity is generated in a gas-fired power station.
   Describe the useful energy transfers that occur.
3) Describe how the following renewable resources are used to generate electricity.
   State one advantage and one disadvantage for each resource.
   a) wind b) solar energy c) the tide d) waves e) geothermal energy
4) Why are hydroelectric power stations often located in remote valleys?
5) What is the purpose of pumped storage?
6) Why is wave power only a realistic major energy source on small islands?
7) What is the source of energy for tidal barrages?
8) Apart from generating electricity, how else can geothermal heat be used?
9) How are biofuels produced? Give two examples of biofuels.
10) Name two places that carbon dioxide can be stored after carbon capture.
11) Name six factors that should be considered when a new power station is being planned.
12) Which three energy sources are linked most strongly with habitat disruption?
13) Explain why a very high electrical voltage is used to transmit electricity in the National Grid.
14) Draw a diagram to illustrate frequency, wavelength and amplitude.
15)* Find the speed of a wave with frequency 50 kHz and wavelength 0.3 cm.
16) a) Sketch a diagram of a ray of light being reflected in a mirror.
    b) Label the normal and the angles of incidence and reflection.
17) Why does light bend as it moves between air and water?
18) Draw a diagram showing a wave diffracting through a gap.
19) What size should the gap be in order to maximise diffraction?
    a) much larger than the wavelength b) the same size as the wavelength c) a bit bigger than the wavelength
20) Sketch the EM spectrum with all its details. Put the lowest frequency waves on the left.
21) What type of wave do television remotes usually use?
22) Which two types of EM wave are commonly used to send signals along optical fibres?
23) Why can't sound waves travel in space?
24) Are high frequency sound waves high pitched or low pitched?
25) If a wave source is moving towards you, will the observed frequency of its waves be higher or lower than their actual frequency?
26) What do red-shift observations tell us about the universe?
27) Describe the 'Big Bang' theory for the origin of the universe. What evidence is there for this theory?

* Answers on p.108.

Velocity and Distance-Time Graphs

Ah, time for some lovely physics, you lucky thing. First off — velocity. The important thing to remember is that if something has velocity it has both speed and direction. Like you, speeding towards success...

Speed and Velocity are Both How Fast You're Going

Speed and velocity are both measured in m/s (or km/h or mph). They both simply say how fast you're going, but there's a subtle difference between them which you need to know:

Speed is just how fast you're going (e.g. 30 mph or 20 m/s) with no regard to the direction.
Velocity however must also have the direction specified, e.g. 30 mph north or 20 m/s, 060°.

Seems kinda fussy I know, but they expect you to remember that distinction, so there you go.

Distance-Time Graphs

These are a very nifty way of describing something travelling through time and space:

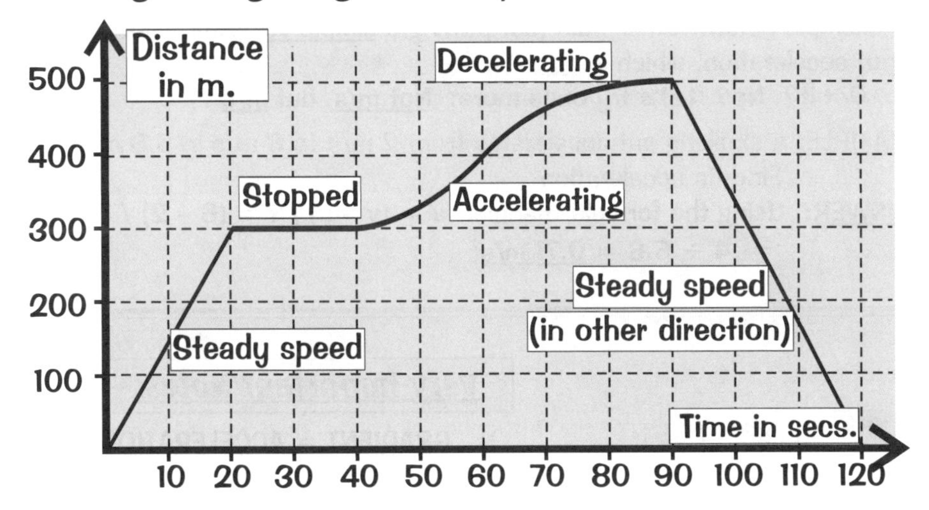

As you probably know, speed = distance ÷ time. So the gradient of a distance-time graph tells you how fast your object is travelling. This is because the gradient is the change in the distance (vertical axis) divided by the change in time (horizontal axis). See — its easy when you know how.

Very Important Notes:

1) Gradient = speed.
2) Flat sections are where it's stationary — it's stopped.
3) Straight uphill or downhill sections mean it is travelling at a steady speed.
4) The steeper the graph, the faster it's going.
5) Downhill sections mean it's going back toward its starting point.
6) Curves represent acceleration or deceleration.
7) A steepening curve means it's speeding up (increasing gradient).
8) A levelling off curve means it's slowing down (decreasing gradient).

Curves = difficulty getting out of chairs.

Calculating Speed from a Distance-Time Graph — It's Just the Gradient

For example the speed of the return section of the graph is:

$$\text{Speed} = \text{gradient} = \frac{\text{vertical}}{\text{horizontal}} = \frac{500}{30} = 16.7 \text{ m/s}$$

Don't forget that you have to use the scales of the axes to work out the gradient. Don't measure in cm!

Ah, speed equals distance over time — that old chestnut...

Distance-time graphs have an annoying habit of popping up in exams year after year — so make sure you're confident with drawing and interpreting them. Remember that the gradient of a distance-time graph is the speed — so the steeper the line, the faster you're going. See — it's simple when you know how.

Acceleration and Velocity-Time Graphs

I bet you loved that distance-time graph, huh? Well here's its big brother — the velocity-time graph. Yikes.

Acceleration is How Quickly Velocity is Changing

Acceleration is definitely not the same as velocity or speed.

1) Acceleration is how quickly the velocity is changing.
2) This change in velocity can be a CHANGE IN SPEED or a CHANGE IN DIRECTION or both.

(You only have to worry about the change in speed bit for calculations.)

Acceleration — The Formula:

$$\text{Acceleration} = \frac{\text{Change in Velocity}}{\text{Time taken}}$$

Well, it's just another formula.
And it's got a formula triangle like all the others.
Mind you, there are two tricky things with this one. First there's the '(v – u)', which means working out the 'change in velocity', as shown in the example below, rather than just putting a simple value for velocity or speed in. Secondly there's the unit of acceleration, which is m/s^2.
Not m/s, which is velocity, but m/s^2. Got it? No? Let's try once more: Not m/s, but m/s^2.

Acceleration is the change in velocity (m/s) per second (s), = m/s^2.

EXAMPLE: A skulking cat accelerates from 2 m/s to 6 m/s in 5.6 s. Find its acceleration.

ANSWER: Using the formula triangle: $a = (v - u) / t = (6 - 2) / 5.6$
$= 4 \div 5.6 = 0.71\ m/s^2$

Velocity-Time Graphs

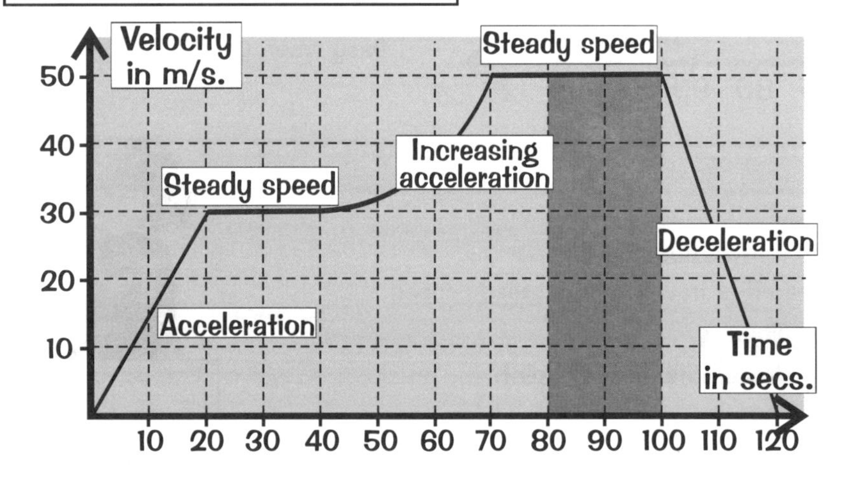

Very Important Notes:

1) GRADIENT = ACCELERATION.
2) Flat sections represent steady speed.
3) The steeper the graph, the greater the acceleration or deceleration.
4) Uphill sections (/) are acceleration.
5) Downhill sections (\) are deceleration.
6) The area under any section of the graph (or all of it) is equal to the distance travelled in that time interval.
7) A curve means changing acceleration.

Calculating Acceleration, Velocity and Distance from a Velocity-Time Graph

1) The acceleration represented by the first section of the graph is:

$$\text{Acceleration} = \text{gradient} = \frac{\text{vertical change}}{\text{horizontal change}} = \frac{30}{20} = 1.5\ m/s^2$$

2) The velocity at any point is simply found by reading the value off the velocity axis.
3) The distance travelled in any time interval is equal to the area under the graph. For example, the distance travelled between t = 80 s and t = 100 s is equal to the shaded area, which is equal to $20 \times 50 = 1000$ m.

Understanding motion graphs — it can be a real uphill struggle...

Make sure you know all there is to know about velocity-time graphs — i.e. learn those numbered points.
You work out acceleration from the graph simply by applying the acceleration formula — change in velocity is the change on the vertical axis and time taken is the change on the horizontal axis.

Weight, Mass and Gravity

Now for something a bit more attractive — the force of gravity. Enjoy...

Gravitational Force is the Force of Attraction Between All Masses

Gravity attracts all masses, but you only notice it when one of the masses is really really big, e.g. a planet. Anything near a planet or star is attracted to it very strongly.

This has two important effects:

1) On the surface of a planet, it makes all things accelerate (see p.44) towards the ground (all with the same acceleration, g, which is about 10 m/s^2 on Earth).
2) It gives everything a weight.

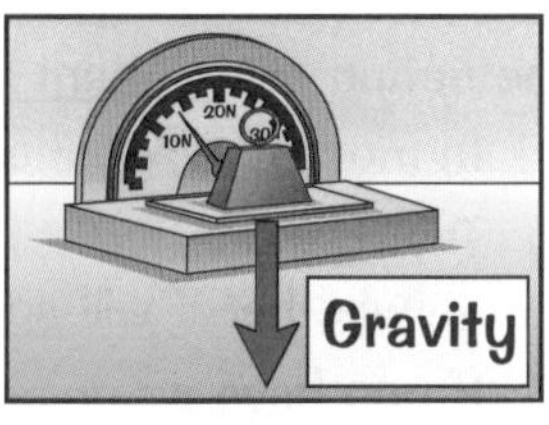

Weight and Mass are Not the Same

1) Mass is just the amount of 'stuff' in an object. For any given object this will have the same value anywhere in the universe.
2) Weight is caused by the pull of the gravitational force. In most questions the weight of an object is just the force of gravity pulling it towards the centre of the Earth.
3) An object has the same mass whether it's on Earth or on the Moon — but its weight will be different. A 1 kg mass will weigh less on the Moon (about 1.6 N) than it does on Earth (about 10 N), simply because the gravitational force pulling on it is less.
4) Weight is a force measured in newtons. It's measured using a spring balance or newton meter. Mass is not a force. It's measured in kilograms with a mass balance (an old-fashioned pair of balancing scales).

The Very Important Formula Relating Mass, Weight and Gravity

weight = mass × gravitational field strength

The acceleration due to gravity and the gravitational field strength are always the same value, no matter what planet or moon you're on.

1) Remember, weight and mass are not the same. Mass is in kg, weight is in newtons.
2) The letter "g" represents the strength of the gravity and its value is different for different planets. On Earth g ≈ 10 N/kg. On the Moon, where the gravity is weaker, g is only about 1.6 N/kg.
3) This formula is hideously easy to use:

Example: What is the weight, in newtons, of a 5 kg mass, both on Earth and on the Moon?

Answer: "$W = m \times g$". On Earth: $W = 5 \times 10 = 50$ N (The weight of the 5 kg mass is 50 N.)
On the Moon: $W = 5 \times 1.6 = 8$ N (The weight of the 5 kg mass is 8 N.)

See what I mean. Hideously easy — as long as you've learnt what all the letters mean.

I don't think you understand the gravity of this situation...

The difference between weight and mass can be tricky to get your head around, but it's well important. Weight is the force of gravity acting on a mass, and mass is the amount of stuff, measured in kg. Now might be a good time to get that equation memorised as well — that's right — cover, scribble and check.

Resultant Forces

Gravity isn't the only force in town — there are other forces such as driving forces or air resistance. What you need to be able to work out is how all these forces add up together.

Resultant Force is the Overall Force on a Point or Object

The notion of resultant force is a really important one for you to get your head round:

1) In most real situations there are at least two forces acting on an object along any direction.
2) The overall effect of these forces will decide the motion of the object — whether it will accelerate, decelerate or stay at a steady speed.
3) If you have a number of forces acting at a single point, you can replace them with a single force (so long as the single force has the same effect on the motion as the original forces acting all together).
4) If the forces all act along the same line (they're all parallel and act in the same or the opposite direction), the overall effect is found by just adding or subtracting them.
5) The overall force you get is called the resultant force.

Example: Stationary Teapot — All Forces Balance

1) The force of GRAVITY (or weight) is acting downwards.
2) This causes a REACTION FORCE (see p.48) from the surface pushing up on the object.
3) This is the only way it can be in BALANCE.
4) Without a reaction force, it would accelerate downwards due to the pull of gravity.
5) The resultant force on the teapot is zero: 10 N – 10 N = 0 N.

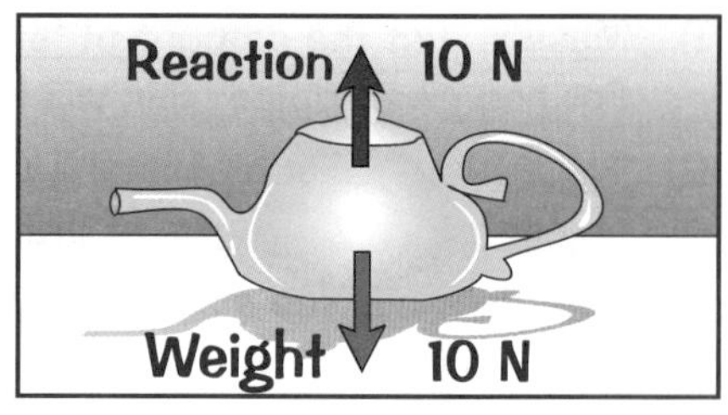

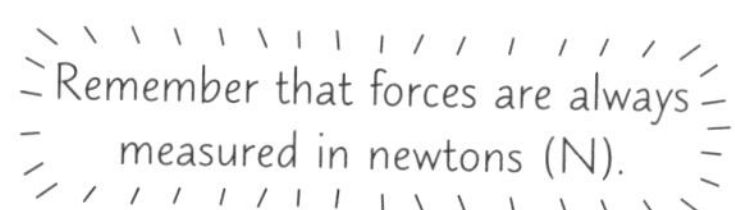

A Resultant Force Means a Change in Velocity

1) If there is a resultant force acting on an object, then the object will change its state of rest or motion.
2) In other words it causes a change in the object's velocity.

You Should be Able to Find the Resultant Force Acting in a Straight Line

EXAMPLE: Benny is cruising along to Las Vegas in his vintage sports car. He applies a driving force of 1000 N, but has to overcome air resistance of 600 N. What is the resultant force? Will the car's velocity change?

ANSWER: Say that the forces pointing to the left are pointing in the positive direction.
The resultant force = 1000 N – 600 N = 400 N to the left.
If there is a resultant force then there is always an acceleration, so Benny's velocity will change. Viva Las Vegas.

And you're moving forward — what a result...

Resultant forces are just about adding and subtracting really — the trick is to make sure you've accounted for everything. Next up, some of the thrilling physics you can understand once you have resultant forces figured out.

Forces and Acceleration

Around about the time of the Great Plague in the 1660s, a chap called Isaac Newton worked out his Laws of Motion. At first they might seem kind of obscure or irrelevant, but to be perfectly blunt, if you can't understand this page then you'll never understand forces and motion.

An Object Needs a Force to Start Moving

If the resultant force on a stationary object is zero, the object will remain stationary.

Things don't just start moving on their own, there has to be a resultant force (see p.46) to get them started.

No Resultant Force Means No Change in Velocity

If there is no resultant force on a moving object it'll just carry on moving at the same velocity.

1) When a train or car or bus or anything else is moving at a constant velocity then the forces on it must all be balanced.
2) Never let yourself entertain the ridiculous idea that things need a constant overall force to keep them moving — NO NO NO NO NO NO!
3) To keep going at a steady speed, there must be zero resultant force — and don't you forget it.

A Resultant Force Means Acceleration

If there is a non-zero resultant force, then the object will accelerate in the direction of the force.

1) A non-zero resultant force will always produce acceleration (or deceleration).
2) This "acceleration" can take five different forms: Starting, stopping, speeding up, slowing down and changing direction.
3) On a force diagram, the arrows will be unequal:

Don't ever say: "If something's moving there must be an overall resultant force acting on it".
Not so. If there's an overall force it will always accelerate.
You get steady speed when there is zero resultant force.
I wonder how many times I need to say that same thing before you remember it?

Steady Speed Bus Tours Ltd. — providing consistent service since 1926...

Objects in space don't need a driving force to keep travelling at a steady speed — it's only because of air resistance and friction that we do. A steady speed means that there is zero resultant force.

Forces and Acceleration

More fun stuff on forces and acceleration here. The big equation to learn is F = ma — it's a really important one and you will be tested on it. Remember that the F is always the resultant force — that's important too.

A Non-Zero Resultant Force Produces an Acceleration

Any resultant force will produce acceleration, and this is the formula for it:

$$F = ma \quad \text{or} \quad a = F/m$$

m = mass in kilograms (kg)
a = acceleration in metres per second squared (m/s^2)
F is the resultant force in newtons (N)

EXAMPLE: A car of mass of 1750 kg has an engine which provides a driving force of 5200 N. At 70 mph the drag force acting on the car is 5150 N. Find its acceleration a) when first setting off from rest b) at 70 mph.

ANSWER: 1) First draw a force diagram for both cases (no need to show the vertical forces):

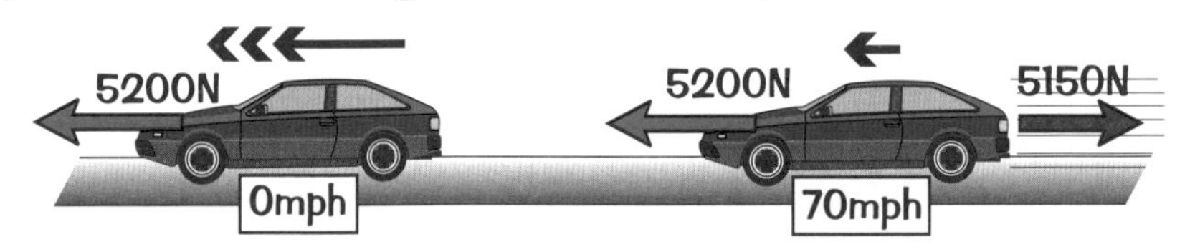

2) Work out the resultant force and acceleration of the car in each case.

Resultant force = 5200 N
a = F/m = 5200 ÷ 1750 = 3.0 m/s^2

Resultant force = 5200 – 5150 = 50 N
a = F/m = 50 ÷ 1750 = 0.03 m/s^2

Reaction Forces are Equal and Opposite

When two objects interact, the forces they exert on each other are equal and opposite.

1) That means if you push something, say a shopping trolley, the trolley will push back against you, just as hard.
2) And as soon as you stop pushing, so does the trolley. Kinda clever really.
3) So far so good. The slightly tricky thing to get your head round is this — if the forces are always equal, how does anything ever go anywhere? The important thing to remember is that the two forces are acting on different objects. Think about a pair of ice skaters:

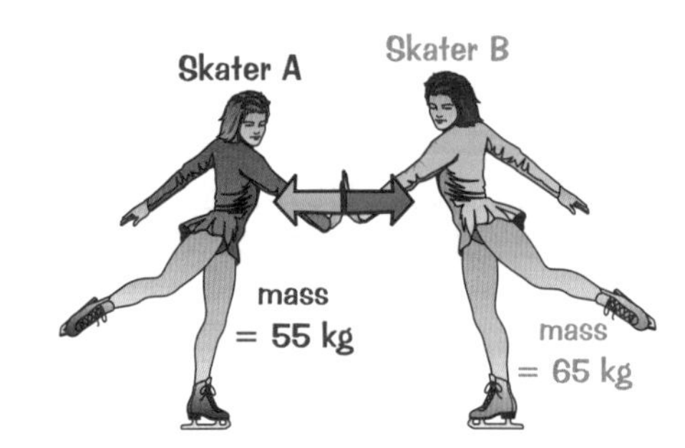

When skater A pushes on skater B (the 'action' force), she feels an equal and opposite force from skater B's hand (the 'reaction' force). Both skaters feel the same sized force, in opposite directions, and so accelerate away from each other.

Skater A will be accelerated more than skater B, though, because she has a smaller mass — remember a = F/m.

4) It's the same sort of thing when you go swimming. You push back against the water with your arms and legs, and the water pushes you forwards with an equal-sized force in the opposite direction.

I have a reaction to forces — they bring me out in a rash...

This is the real deal. Like... proper Physics. It was pretty fantastic at the time it was discovered — suddenly people understood how forces and motion worked, they could work out the orbits of planets and everything. Inspired? No? Shame. Learn it anyway — you're really going to struggle in the exam if you don't.

Frictional Force and Terminal Velocity

Ever wondered why it's so hard to run into a hurricane whilst wearing a sandwich board? Read on to find out...

Friction is Always There to Slow Things Down

1) If an object has no force propelling it along it will always slow down and stop because of friction (unless you're in space where there's nothing to rub against).
2) Friction always acts in the opposite direction to movement.
3) To travel at a steady speed, the driving force needs to balance the frictional forces.
4) You get friction between two surfaces in contact, or when an object passes through a fluid (drag).

RESISTANCE OR "DRAG" FROM FLUIDS (air or liquid)

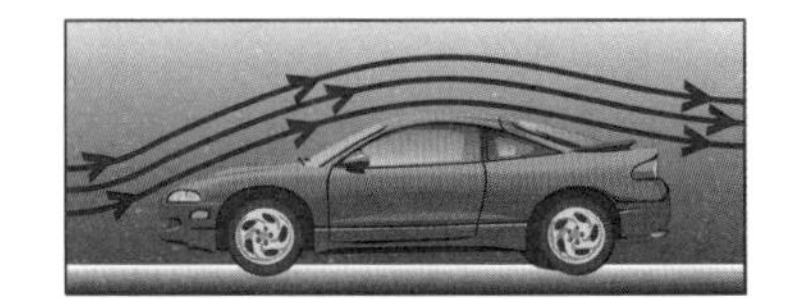

Most of the resistive forces are caused by air resistance or "drag". The most important factor by far in reducing drag in fluids is keeping the shape of the object streamlined. The opposite extreme is a parachute which is about as high drag as you can get — which is, of course, the whole idea.

Drag Increases as the Speed Increases

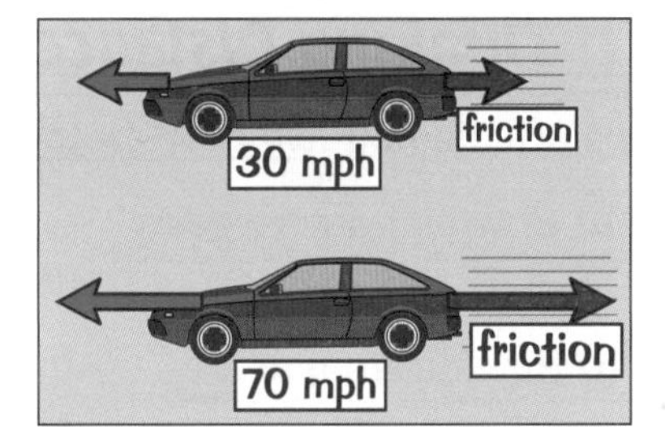

Frictional forces from fluids always increase with speed. A car has much more friction to work against when travelling at 70 mph compared to 30 mph. So at 70 mph the engine has to work much harder just to maintain a steady speed.

Objects Falling Through Fluids Reach a Terminal Velocity

When falling objects first set off, the force of gravity is much more than the frictional force slowing them down, so they accelerate. As the speed increases the friction builds up. This gradually reduces the acceleration until eventually the frictional force is equal to the accelerating force and then it won't accelerate any more. It will have reached its maximum speed or terminal velocity and will fall at a steady speed.

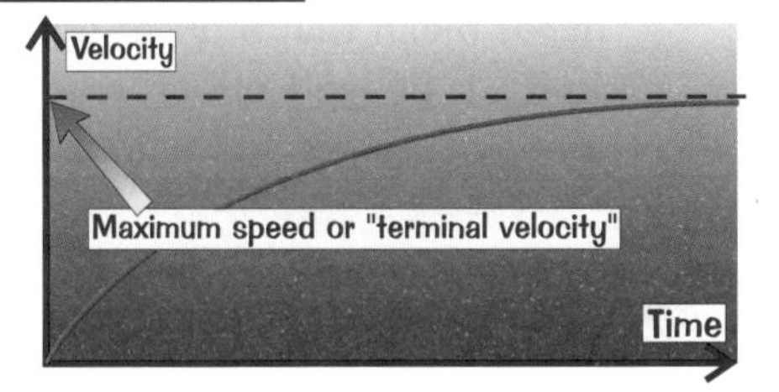

The Terminal Velocity of Falling Objects Depends on their Shape and Area

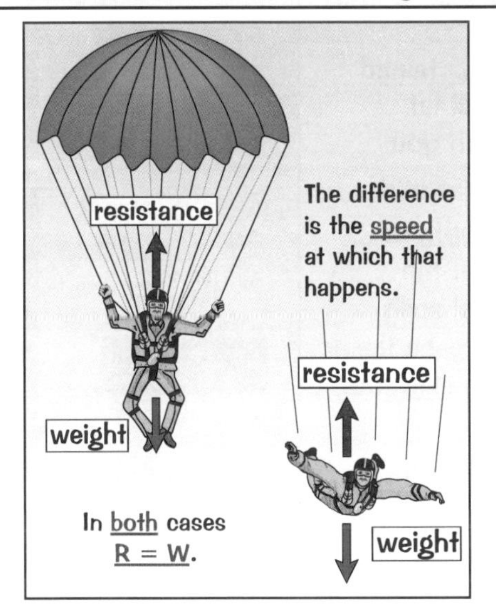

The accelerating force acting on all falling objects is gravity and it would make them all fall at the same rate, if it wasn't for air resistance. This means that on the Moon, where there's no air, hamsters and feathers dropped simultaneously will hit the ground together. However, on Earth, air resistance causes things to fall at different speeds, and the terminal velocity of any object is determined by its drag in comparison to its weight. The frictional force depends on its shape and area.

The most important example is the human skydiver. Without his parachute open he has quite a small area and a force of "$W = mg$" pulling him down. He reaches a terminal velocity of about 120 mph. But with the parachute open, there's much more air resistance (at any given speed) and still only the same force "$W = mg$" pulling him down. This means his terminal velocity comes right down to about 15 mph, which is a safe speed to hit the ground at.

Learning about air resistance — it can be a real drag...

There are a few really important things on this page. 1) When you fall through a fluid, there's a frictional force (drag), 2) frictional force increases with speed, so 3) you eventually reach terminal velocity.

Stopping Distances

And now a page on stopping distances. This may seem a bit out of kilter with the rest of the section, but it's a real world application of the physics of forces. See, I told you it was useful... and fun... right?

Many Factors Affect Your Total Stopping Distance

1) Looking at things simply — if you need to stop in a given distance, then the faster a vehicle's going, the bigger braking force it'll need.
2) Likewise, for any given braking force, the faster you're going, the greater your stopping distance. But in real life it's not quite that simple — if your maximum braking force isn't enough, you'll go further before you stop.
3) The total stopping distance of a vehicle is the distance covered in the time between the driver first spotting a hazard and the vehicle coming to a complete stop.
4) The stopping distance is the sum of the thinking distance and the braking distance.

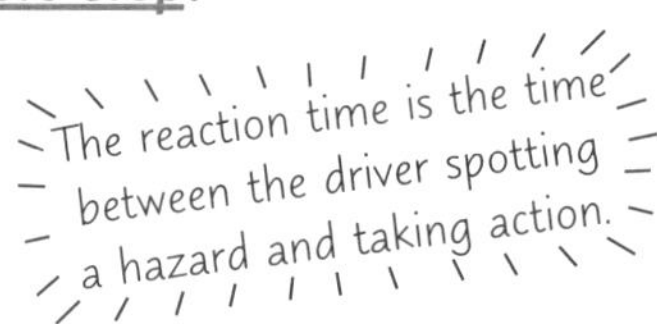

1) Thinking Distance

"The distance the vehicle travels during the driver's reaction time".

It's affected by two main factors:

a) How fast you're going — Obviously. Whatever your reaction time, the faster you're going, the further you'll go.

b) How dopey you are — This is affected by tiredness, drugs, alcohol and a careless blasé attitude.

Bad visibility and distractions can also be a major factor in accidents — lashing rain, messing about with the radio, bright oncoming lights, etc. might mean that a driver doesn't notice a hazard until they're quite close to it. It doesn't affect your thinking distance, but you start thinking about stopping nearer to the hazard, and so you're more likely to crash.

The figures below for typical stopping distances are from the Highway Code. It's frightening to see just how far it takes to stop when you're going at 70 mph.

30 mph
9 m
14 m
6 car lengths

50 mph
15 m
38 m
13 car lengths

70 mph
21 m
75 m
24 car lengths

Thinking distance

Braking distance

2) Braking Distance

"The distance the car travels under the breaking force".

It's affected by four main factors:

a) How fast you're going — The faster you're going, the further it takes to stop.

b) How good your brakes are — All brakes must be checked and maintained regularly. Worn or faulty brakes will let you down catastrophically just when you need them the most, i.e. in an emergency.

c) How good the tyres are — Tyres should have a minimum tread depth of 1.6 mm in order to be able to get rid of the water in wet conditions. Leaves, diesel spills and muck on the road can greatly increase the braking distance, and cause the car to skid too.

d) How good the grip is — This depends on three things: 1) road surface, 2) weather conditions, 3) tyres.

Wet or icy roads are always much more slippy than dry roads, but often you only discover this when you try to brake hard. You don't have as much grip, so you travel further before stopping.

Stop right there — and learn this page...

Without tread, a tyre will simply ride on a layer of water and skid very easily. This is called "aquaplaning" and isn't nearly as cool as it sounds. Snow and ice are also very hazardous because it is difficult for the tyres to get a grip.

Work and Potential Energy

When a force moves an object through a distance, ENERGY IS TRANSFERRED and WORK IS DONE.

That statement sounds far more complicated than it needs to. Try this:

1) Whenever something moves, something else is providing some sort of 'effort' to move it.
2) The thing putting the effort in needs a supply of energy (like fuel or food or electricity etc.).
3) It then does 'work' by moving the object — and one way or another it transfers the energy it receives (as fuel) into other forms.

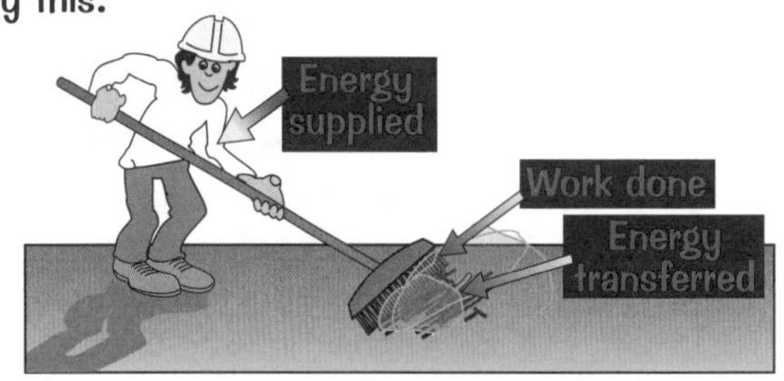

4) Whether this energy is transferred 'usefully' (e.g. by lifting a load) or is 'wasted' (e.g. lost as heat through friction), you can still say that 'work is done'. Just like Batman and Bruce Wayne, 'work done' and 'energy transferred' are indeed 'one and the same'. (And they're both given in joules.)

It's Just Another Trivial Formula:

Work Done = Force × Distance

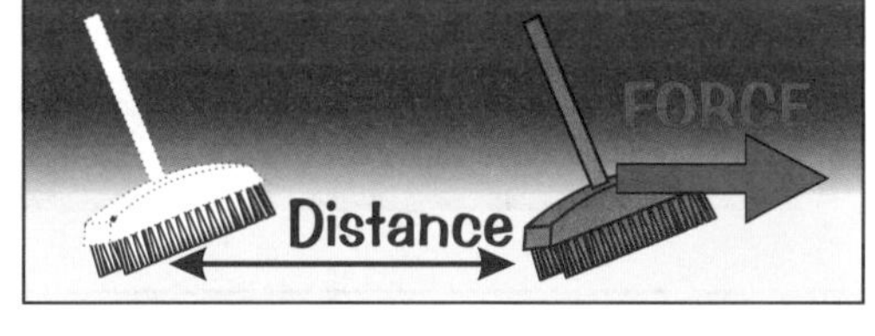

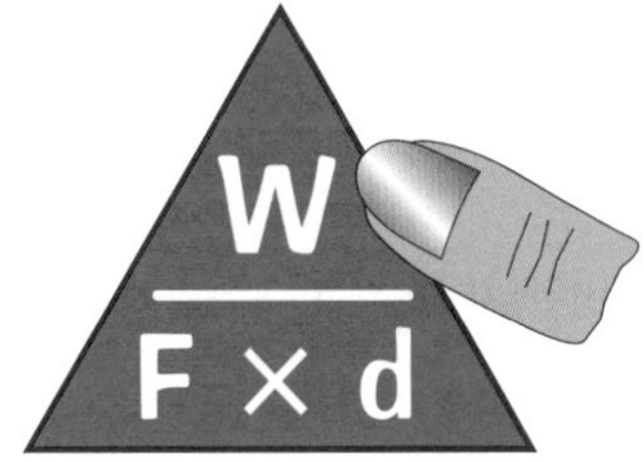

Whether the force is friction or weight or tension in a rope, it's always the same. To find how much energy has been transferred (in joules), you just multiply the force in N by the distance moved in m. Easy as that. I'll show you...

EXAMPLE: Some hooligan kids drag an old tractor tyre 5 m over rough ground. They pull with a total force of 340 N. Find the energy transferred.

ANSWER: $W = F \times d = 340 \times 5 = 1700$ J. Phew — easy peasy isn't it?

Gravitational Potential Energy is Energy Due to Height

Gravitational Potential Energy = mass × g × height

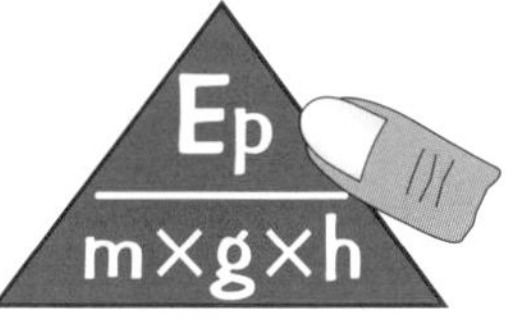

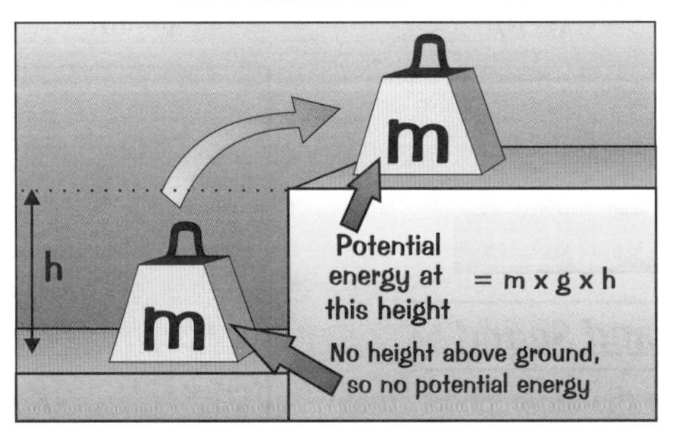

Gravitational potential energy (measured in joules) is the energy that an object has by virtue of (because of) its vertical position in a gravitational field. When an object is raised vertically, work is done against the force of gravity (it takes effort to lift it up) and the object gains gravitational potential energy. On Earth the gravitational field strength (g) is approximately 10 N/kg.

EXAMPLE: A sheep of mass 47 kg is slowly raised through 6.3 m. Find the gain in potential energy.

ANSWER: Just plug the numbers into the formula:
$E_p = mgh = 47 \times 10 \times 6.3 = 2961$ J
(Joules because it's energy.)

Revise work done — what else...

Remember "energy transferred" and "work done" are the same thing. By lifting something up you do work by transferring chemical energy to gravitational potential energy. Think about that next time you're bench-pressing sheep.

Kinetic Energy

Kinetic Energy is Energy of Movement

Anything that's moving has kinetic energy.
There's a slightly tricky formula for it, so you have to concentrate a little bit harder for this one.
But hey, that's life — it can be real tough sometimes:

Kinetic Energy = ½ × mass × speed²

$E_k = \frac{1}{2} \times m \times v^2$

EXAMPLE: A car of mass 2450 kg is travelling at 38 m/s.
Calculate its kinetic energy.

ANSWER: It's pretty easy. You just plug the numbers into the formula — but watch the 'v^2'!
K.E. = $\frac{1}{2}mv^2 = \frac{1}{2} \times 2450 \times 38^2$ = 1 768 900 J (Joules because it's energy.)

Remember, the kinetic energy of something depends both on mass and speed.
The more it weighs and the faster it's going, the bigger its kinetic energy will be.

Kinetic Energy Transferred is Work Done

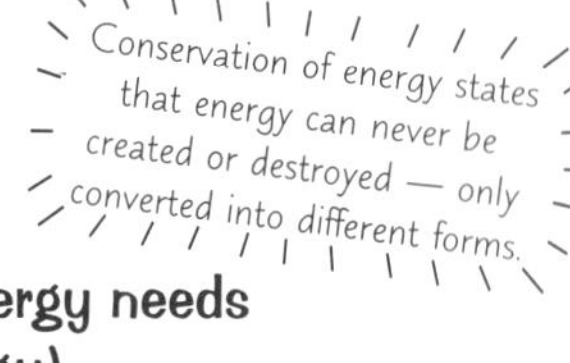

When a Car is Moving It Has Kinetic Energy

1) A moving car can have a lot of kinetic energy. To slow a car down this kinetic energy needs to be converted into other types of energy (using the law of conservation of energy).
2) To stop a car, the kinetic energy ($\frac{1}{2}mv^2$) has to be converted to heat energy as friction between the wheels and the brake pads, causing the temperature of the brakes to increase:

Kinetic Energy Transferred = Work Done by Brakes

$$\frac{1}{2}mv^2 = F \times d$$

m = mass of car and passengers (in kg) v = speed of car (in m/s) F = maximum braking force (in N) d = braking distance (in m).

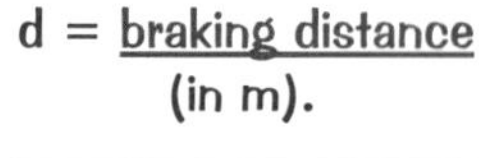

Falling Objects Convert E_P into E_K...

When something falls, its potential energy (see p. 51) is converted into kinetic energy. So the further it falls, the faster it goes.

Kinetic energy gained = Potential Energy lost

P.E. → K.E.

...and some of this E_K is Transferred into Heat and Sound

When meteors and space shuttles enter the atmosphere, they have a very high kinetic energy. Friction due to collisions with particles in the atmosphere transfers some of their kinetic energy to heat energy and work is done. The temperatures can become so extreme that most meteors burn up completely and never hit the Earth. Only the biggest meteors make it through to the Earth's surface — these are called meteorites.

Space shuttles have heat shields made from special materials which lose heat quickly, allowing the shuttle to re-enter the atmosphere without burning up.

Kinetic energy — just get a move on and learn it, OK...

So that's why I've not been hit by a meteor — most get burned up. Now you know. What you probably don't know yet, though, is that rather lovely formula at the top of the page. I mean, gosh, it's got more than three letters in it...

Forces and Elasticity

Forces aren't just important for cars and falling sheep — you can stretch things with them as well. It can sound quite tricky at first, but it's not as hard as it looks. And there is only one equation to memorise — hurrah.

Work Done to an Elastic Object is Stored as Elastic Potential Energy

1) When you apply a force to an object you may cause it to stretch and change in shape.
2) Any object that can go back to its original shape after the force has been removed is an elastic object.
3) Work is done to an elastic object to change its shape. This energy is not lost but is stored by the object as elastic potential energy.
4) The elastic potential energy is then converted to kinetic energy when the force is removed and the object returns to its original shape, e.g. when a spring or an elastic band bounces back.

Elastic potential energy — useful for passing exams and scaring small children

Extension of an Elastic Object is Directly Proportional to Force...

If a spring is supported at the top and then a weight attached to the bottom, it stretches.

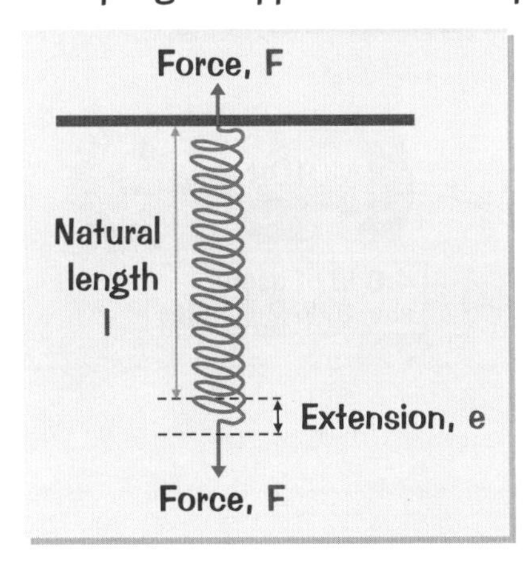

1) The extension, e, of a stretched spring (or other elastic object) is directly proportional to the load or force applied, F. The extension is measured in metres, and the force is measured in newtons.
2) This is the equation you need to learn:

3) k is the spring constant. Its value depends on the material that you are stretching and it's measured in newtons per metre (N/m).

...but this Stops Working when the Force is Great Enough

There's a limit to the amount of force you can apply to an object for the extension to keep on increasing proportionally.

1) The graph shows force against extension for an elastic object.
2) For small forces, force and extension are proportional. So the first part of the graph shows a straight-line relationship between force and extension.
3) There is a maximum force that the elastic object can take and still extend proportionally. This is known as the limit of proportionality and is shown on the graph at the point marked P.
4) If you increase the force past the limit of proportionality, the material will be permanently stretched. When the force is removed, the material will be longer than at the start.

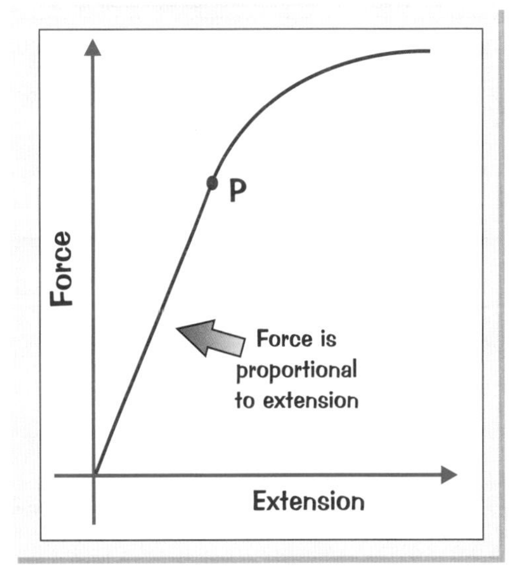

I could make a joke, but I don't want to stretch myself...

Scaring small children aside, elastic potential is really quite a useful form of energy. Think of all the things we rely on that use it — catapults, trampolines, scrunchies... Ah, elastic potential energy — thank you for enriching our lives.

Power

Power is a concept that pops up in both forces and electricity. This is because, at its most fundamental level, power is just about the rate of energy transferred — and energy is transferred wherever you look.

Power is the "Rate of Doing Work" — i.e. How Much per Second

Power is not the same thing as force, nor energy. A powerful machine is not necessarily one which can exert a strong force (though it usually ends up that way).
A powerful machine is one which transfers a lot of energy in a short space of time.
This is the very easy formula for power:

$$\text{Power} = \frac{\text{Work done (or energy transferred)}}{\text{Time taken}}$$

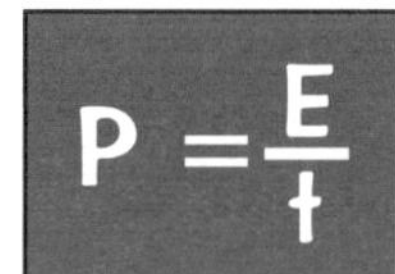

Power is Measured in Watts (or J/s)

The proper unit of power is the watt. One watt = 1 joule of energy transferred per second.
Power means "how much energy per second", so watts are the same as "joules per second" (J/s).
Don't ever say "watts per second" — it's nonsense.

Example: A motor transfers 4.8 kJ of useful energy in 2 minutes. Find its power output.
Answer: $P = E / t = 4800/120 = 40$ W (or 40 J/s)
(Note that the kJ had to be turned into J, and the minutes into seconds.)

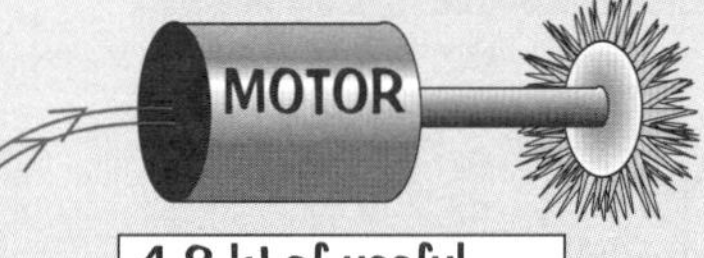

Calculating Your Power Output

There are a few different ways to measure the power output of a person:

a) The Timed Run Upstairs:

In this case the "energy transferred" is the potential energy you gain (= mgh).
Hence Power = mgh/t

Power output
= En. transferred/time
= mgh/t
= $(62 \times 10 \times 12) \div 14$
= 531 W

b) The Timed Acceleration:

This time the energy transferred is the kinetic energy you gain (= $\frac{1}{2}mv^2$).
Hence Power = $\frac{1}{2}mv^2/t$

Power output
= En. transferred/time
= $\frac{1}{2}mv^2/t$
= $(\frac{1}{2} \times 62 \times 8^2) \div 4$
= 496 W

To get accurate results from these experiments, you have to do them several times and find an average.

Power — you need to know watt's watt...

Power is the amount of energy transferred per second, and it's measured in watts. The watt is named after James Watt, a Scottish inventor and engineer who did a lot of work on steam engines in the 1700s. Nice. Make sure you learn the formula and power questions should be a doddle.

Momentum and Collisions

A large rhino running very fast at you is going to be a lot harder to stop than a scrawny one out for a Sunday afternoon stroll — that's momentum for you.

Momentum = Mass × Velocity

1) Momentum (p) is a property of moving objects.
2) The greater the mass of an object and the greater its velocity (see p. 43) the more momentum the object has.
3) Momentum is a vector quantity — it has size and direction (like velocity, but not speed).

$p = m \times v$

Momentum (kg m/s) = Mass (kg) × Velocity (m/s)

Momentum Before = Momentum After

In a closed system, the total momentum before an event (e.g. a collision) is the same as after the event. This is called Conservation of Momentum.

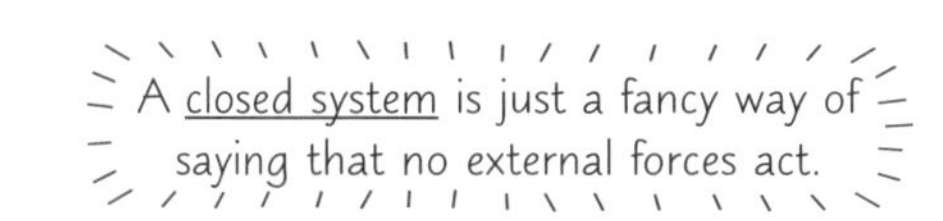

Example 1: Collisions

Two skaters approach each other, collide and move off together as shown. At what velocity do they move after the collision?

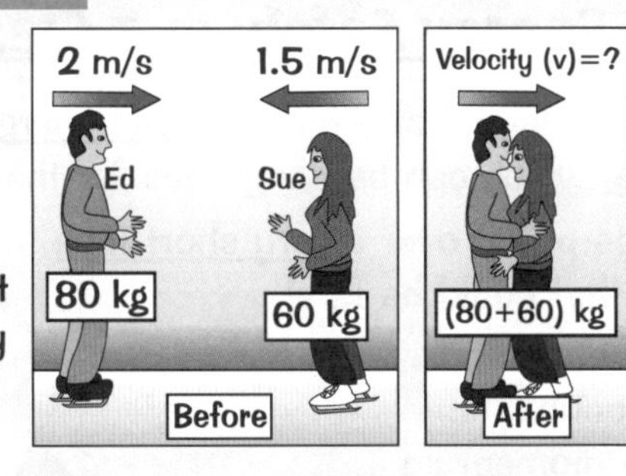

1) Choose which direction is positive.
   I'll say "positive" means "to the right".
2) Total momentum before collision
   = momentum of Ed + momentum of Sue
   = {80 × 2} + {60 × (–1.5)}
   = 70 kg m/s
3) Total momentum after collision
   = momentum of Ed and Sue together
   = 140 × v
4) So 140v = 70, i.e. v = 0.5 m/s to the right

Example 2: Explosions

A gun fires a bullet as shown. At what speed does the gun move backwards?

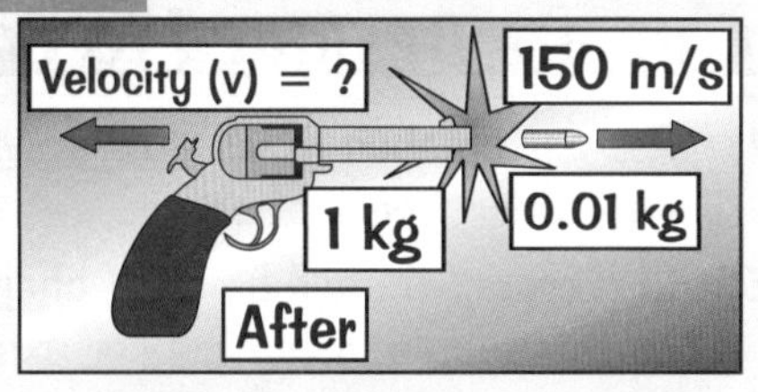

1) Choose which direction is positive.
   Again, I reckon "positive" means "to the right".
2) Total momentum before firing = 0 kg m/s
3) Total momentum after firing
   = momentum of bullet + momentum of gun
   = (0.01 × 150) + (1 × v)
   = 1.5 + v

   This is the gun's recoil.

   The momentum of a system before an explosion is zero, so, due to conservation of momentum, the total momentum after an explosion is zero too.
4) So 1.5 + v = 0, i.e. v = –1.5 m/s
   So the gun moves backwards at 1.5 m/s.

Forces Cause Changes in Momentum

1) When a force acts on an object, it causes a change in momentum.
2) A larger force means a faster change of momentum (and so a greater acceleration).
3) Likewise, if someone's momentum changes very quickly (like in a car crash), the forces on the body will be very large, and more likely to cause injury.
4) This is why cars are designed with safety features that slow people down over a longer time when they have a crash — the longer it takes for a change in momentum, the smaller the force.

Learn this stuff — it'll only take a moment... um...

Momentum's a pretty fundamental bit of Physics — so make sure you learn it properly. Right then, momentum is always conserved in collisions and explosions when there are no external forces acting. Job's a good 'un.

Car Design and Safety

A lot of the physics you've seen over the last few pages can be applied in the real world to designing safe and efficient cars. It's all about forces, energy, acceleration and momentum. Sweet as a nut.

Brakes do Work Against the Kinetic Energy of the Car

When you apply the brakes to slow down a car, work is done (see p.51).
The brakes reduce the kinetic energy of the car by transferring it into heat (and sound) energy (see p.52).
In traditional braking systems that would be the end of the story, but new regenerative braking systems used in some electric or hybrid cars make use of the energy, instead of converting it all into heat during braking.

1) Regenerative brakes use the system that drives the vehicle to do the majority of the braking.
2) Rather than converting the kinetic energy of the vehicle into heat energy, the brakes put the vehicle's motor into reverse. With the motor running backwards, the wheels are slowed.
3) At the same time, the motor acts as an electric generator, converting kinetic energy into electrical energy that is stored as chemical energy in the vehicle's battery. This is the advantage of regenerative brakes — they store the energy of braking rather than wasting it. It's a nifty chain of energy transfer.

Cars are Designed to Convert Kinetic Energy Safely in a Crash

1) If a car crashes it will slow down very quickly — this means that a lot of kinetic energy is converted into other forms of energy in a short amount of time, which can be dangerous for the people inside.
2) In a crash, there'll be a big change in momentum (see p.55) over a very short time, so the people inside the car experience huge forces that could be fatal.
3) Cars are designed to convert the kinetic energy of the car and its passengers in a way that is safest for the car's occupants. They often do this by increasing the time over which momentum changes happen, which lessens the forces on the passengers.

airbag

seat belt

CRUMPLE ZONES at the front and back of the car crumple up on impact.

- The car's kinetic energy is converted into other forms of energy by the car body as it changes shape.
- Crumple zones increase the impact time, decreasing the force produced by the change in momentum.

SIDE IMPACT BARS are strong metal tubes fitted into car door panels. They help direct the kinetic energy of the crash away from the passengers to other areas of the car, such as the crumple zones.

SEAT BELTS stretch slightly, increasing the time taken for the wearer to stop. This reduces the forces acting in the chest. Some of the kinetic energy of the wearer is absorbed by the seat belt stretching.

AIR BAGS also slow you down more gradually and prevent you from hitting hard surfaces inside the car.

Cars Have Different Power Ratings

Sports car power = 100 kW

1) The size and design of car engines determine how powerful they are.
2) The more powerful an engine is, the more energy it transfers from its fuel every second, and so the faster its top speed can be.
3) E.g. the power output of a typical small car will be around 50 kW and a sports car will be about 100 kW (some are much higher).

Small car power = 50 kW

4) Cars are also designed to be aerodynamic. This means that they are shaped in such a way that air flows very easily and smoothly past them, so minimising their air resistance.

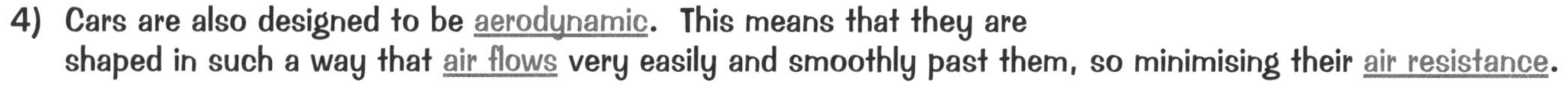

5) Cars reach their top speed when the resistive force equals the driving force provided by the engine (see p.49). So, with less air resistance to overcome, the car can reach a higher speed before this happens. Aerodynamic cars therefore have higher top speeds.

Don't let all this revising drive you crazy...

Driving can be quite risky when you look at the physics of it — which is why so much time and effort is put into making cars as safe as possible. There's a lot of info on this page but sadly you've got to learn it. Sozzles.

Revision Summary for Physics 2a

Well done — you've made it to the end of another section. There are loads of bits and bobs about forces, motion and fast cars which you have to learn. The best way to find out what you know is to get stuck in to these lovely revision questions, which you're going to really enjoy (honest)...

1) What's the difference between speed and velocity?
2) Sketch a typical distance-time graph and point out all the important parts of it.
3) Explain how to calculate speed from a distance-time graph.
4) What is acceleration? What are its units?
5)* Write down the formula for acceleration. What's the acceleration of a soggy pea flicked from rest to a speed of 14 m/s in 0.4 seconds?
6) Sketch a typical velocity-time graph and point out all the important parts of it.
7) Explain how to find speed, distance and acceleration from a velocity-time graph.
8) Explain the difference between mass and weight. What units are they measured in?
9) Explain what is meant by a "resultant force".
10) If an object has zero resultant force on it, can it be moving? Can it be accelerating?
11)* Write down the formula relating resultant force and acceleration. A resultant force of 30 N pushes a trolley of mass 4 kg. What will be its acceleration?
12)* A skydiver has a mass of 75 kg. At 80 mph, the drag force on the skydiver is 650 N. Find the acceleration of the skydiver at 80 mph (take $g = 10$ N/kg).
13)* A yeti pushes a tree with a force of 120 N. What is the size of the reaction force that the Yeti feels pushing back at him?
14) What is "terminal velocity"?
15) What are the two different parts of the overall stopping distance of a car?
16)* Write down the formula for work done. A crazy dog drags a big branch 12 m over the next-door neighbour's front lawn, pulling with a force of 535 N. How much work was done?
17)* A 4 kg cheese is taken 30 m up a hill before being rolled back down again. If $g = 10$ N/kg:
    a) how much gravitational potential energy does the cheese have at the top of the hill?
    b) how much gravitational potential energy does it have when it gets half way down?
18)* What's the formula for kinetic energy? Find the kinetic energy of a 78 kg sheep moving at 23 m/s.
19)* Calculate the kinetic energy of the same 78 kg sheep just as she hits the floor after falling through 20 m.
20)* A car of mass 1000 kg is travelling at a velocity of 2 m/s when a dazed and confused sheep runs out 5 m in front. If the driver immediately applies the maximum braking force of 395 N, can he avoid hitting it?
21) Write down the equation that relates the force on a spring and its extension.
22) What happens to an elastic object that is stretched beyond its limit of proportionality?
23)* Calculate the power output of that 78 kg sheep when she runs 20 m up a staircase in 16.5 seconds.
24) Write down the formula for momentum.
25) If the total momentum of a system before a collision is zero, what is the total momentum of the system after the collision?
26) What is the advantage of using regenerative braking systems?
27) Explain how seat belts, crumple zones, side impact bars and air bags are useful in a crash.
28) Describe the effect on the top speed of a car of adding a roof box. Explain your answer.

* Answers on p.108.

Static Electricity

Static electricity is all about charges which are not free to move, e.g. in insulating materials. This causes them to build up in one place and it often ends with a spark or a shock when they do finally move.

Build-up of Static is Caused by Friction

1) When certain insulating materials are rubbed together, negatively charged electrons will be scraped off one and dumped on the other.
2) This'll leave a positive static charge on one and a negative static charge on the other.
3) Which way the electrons are transferred depends on the two materials involved.
4) Electrically charged objects attract small objects placed near them.
(Try this: rub a balloon on a woolly pully — then put it near tiddly bits of paper and watch them jump.)
5) The classic examples are polythene and acetate rods being rubbed with a cloth duster, as shown in the diagrams.

With the polythene rod, electrons move from the duster to the rod.

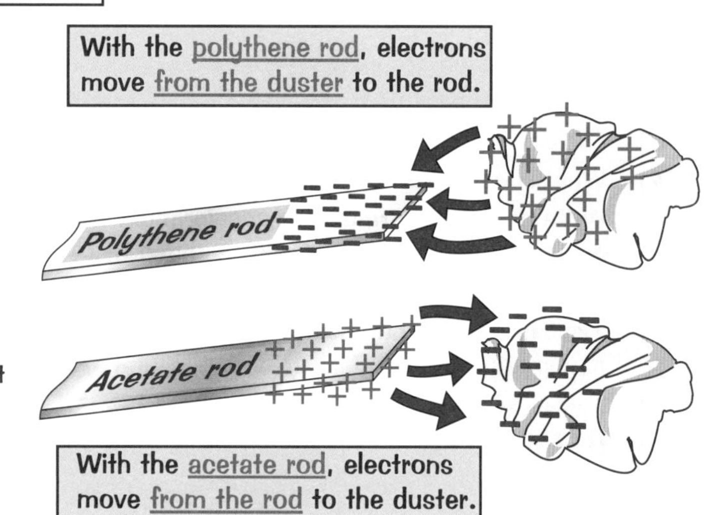

With the acetate rod, electrons move from the rod to the duster.

Only Electrons Move — Never the Positive Charges

Watch out for this in exams. Both +ve and –ve electrostatic charges are only ever produced by the movement of electrons. The positive charges definitely do not move!

A positive static charge is always caused by electrons moving away elsewhere. The material that loses the electrons loses some negative charge, and is left with an equal positive charge, as shown above. Don't forget!

Like Charges Repel, Opposite Charges Attract

This is easy and, I'd have thought, kind of obvious. When two electrically charged objects are brought close together they exert a force on one another.

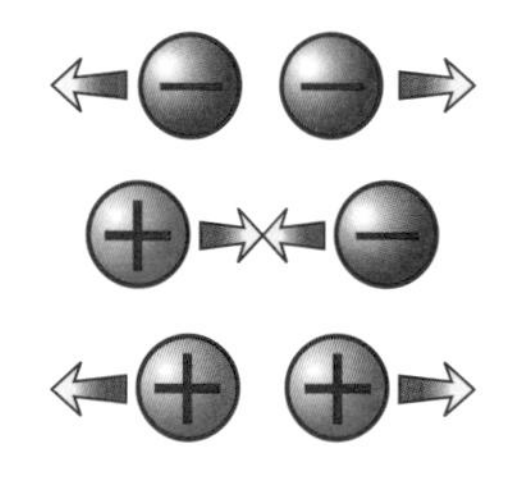

Two things with opposite electric charges are attracted to each other.
Two things with the same electric charge will repel each other.
These forces get weaker the further apart the two things are.

Charges can Move Easily Through Conductors

1) Electrical charges can move easily through some materials. These materials are called conductors.
2) Metals are known to be good conductors.

Stay away from electrons — they're a negative influence...

The bog standard electrical charge carrier is the electron. Those little devils get just about everywhere in metals, taking charge pretty much wherever you want it. But in insulators they're stuck and can't move easily — its only when they're manually scraped off that they ever get to go anywhere, poor blighters...

Current and Potential Difference

Isn't electricity great. Mind you it's pretty bad news if the words don't mean anything to you... Hey, I know — learn them now!

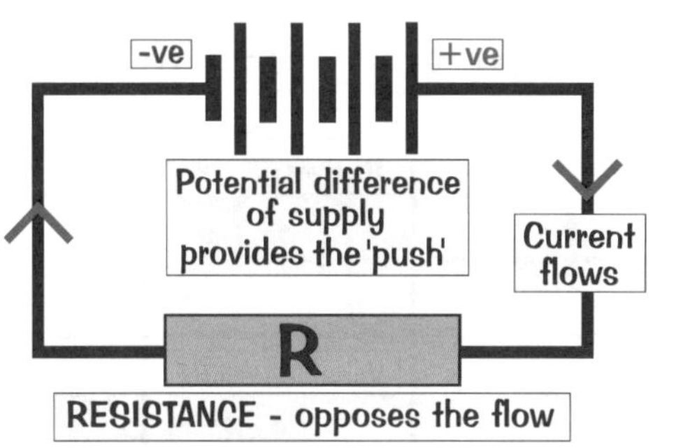

1) Current is the flow of electric charge round the circuit. Current will only flow through a component if there is a potential difference across that component. Unit: ampere, A.

2) Potential Difference is the driving force that pushes the current round. Unit: volt, V.

3) Resistance is anything in the circuit which slows the flow down. Unit: ohm, Ω.

The greater the resistance across a component, the smaller the current that flows (for a given potential difference across the component).

Total Charge Through a Circuit Depends on Current and Time

1) Current is the rate of flow of charge. When current (I) flows past a point in a circuit for a length of time (t) then the charge (Q) that has passed is given by this formula:

2) Current is measured in amperes (A), charge is measured in coulombs (C), time is measured in seconds (s).

$$\text{Current} = \frac{\text{Charge}}{\text{Time}}$$

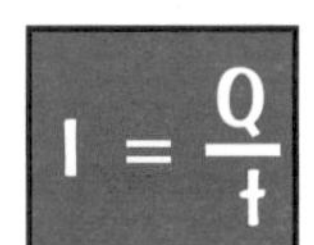

3) More charge passes around the circuit when a bigger current flows.

EXAMPLE: A battery charger passes a current of 2.5 A through a cell over a period of 4 hours. How much charge does the charger transfer to the cell altogether?

ANSWER: $Q = I \times t = 2.5 \times (4 \times 60 \times 60) = 36\,000$ C (36 kC).

Potential Difference (P. D.) is the Work Done Per Unit Charge

1) The potential difference (or voltage) is the work done (the energy transferred, measured in joules, J) per coulomb of charge that passes between two points in an electrical circuit. It's given by this formula:

$$\text{P.D.} = \frac{\text{Work done}}{\text{Charge}}$$

2) So, the potential difference across an electrical component is the amount of energy that is transferred by that electrical component (e.g. to light and heat energy by a bulb) per unit of charge.

3) Voltage and potential difference mean the same thing. You can use either in your exam and scoop up the marks (so long as you use it correctly).

I think it's about time you took charge...

Don't get confused by the words voltage and potential difference — they mean the same thing. Just remember that the potential difference is the work done between two points in a circuit, per unit of charge. Get those two formulas learned as well — examiners just love to test whether you understand them.

Circuits — The Basics

Formulas are mighty pretty and all, but you might have to design some electrical circuits as well one day. For that you're gonna need circuit symbols. Well, would you look at that... they're on this page.

Circuit Symbols You Should Know — Learn Them Well

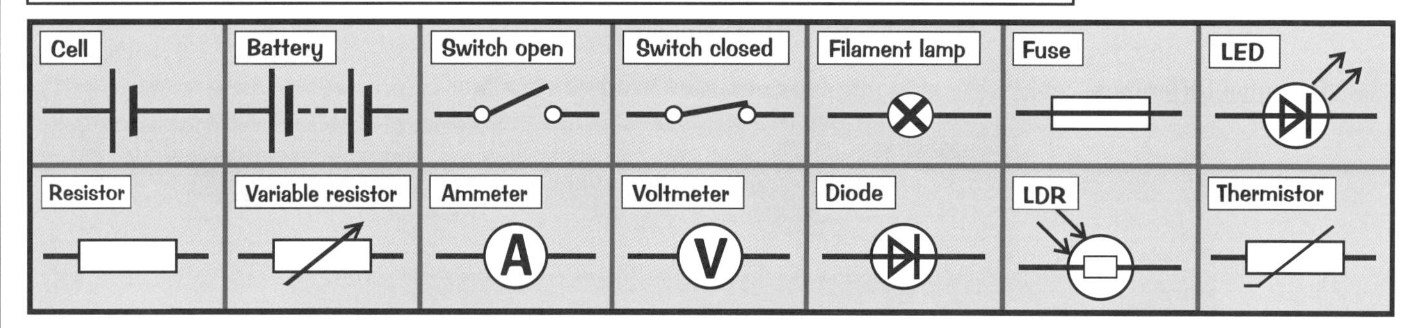

The Standard Test Circuit

This is the circuit you use if you want to know the resistance of a component.
You find the resistance by measuring the current through and the potential difference across the component. It is absolutely the most bog standard circuit you could know. So know it.

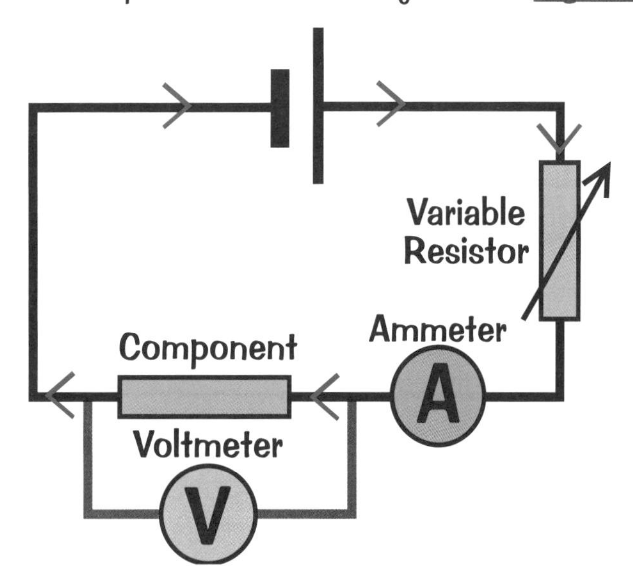

The Ammeter

1) Measures the current (in amps) flowing through the component.
2) Must be placed in series (see p.63).
3) Can be put anywhere in series in the main circuit, but never in parallel like the voltmeter.

The Voltmeter

1) Measures the potential difference (in volts) across the component.
2) Must be placed in parallel (see p.64) around the component under test — NOT around the variable resistor or the battery!

Five Important Points

1) This very basic circuit is used for testing components, and for getting V-I graphs from them (see next page).
2) The component, the ammeter and the variable resistor are all in series, which means they can be put in any order in the main circuit. The voltmeter, on the other hand, can only be placed in parallel around the component under test, as shown. Anywhere else is a definite no-no.
3) As you vary the variable resistor it alters the current flowing through the circuit.
4) This allows you to take several pairs of readings from the ammeter and voltmeter.
5) You can then plot these values for current and voltage on a V-I graph and find the resistance.

Measure gymnastics — use a vaultmeter...

The funny thing is — the electrons in circuits actually move from –ve to +ve... but scientists always think of current as flowing from +ve to –ve. Basically it's just because that's how the early physicists thought of it (before they found out about the electrons), and now it's become convention.

Resistance and $V = I \times R$

With your current and your potential difference measured, you can now make some sweet graphs...

Three Hideously Important Potential Difference-Current Graphs

V-I graphs show how the current varies as you change the potential difference (P.D.). Learn these three real well:

Different Resistors

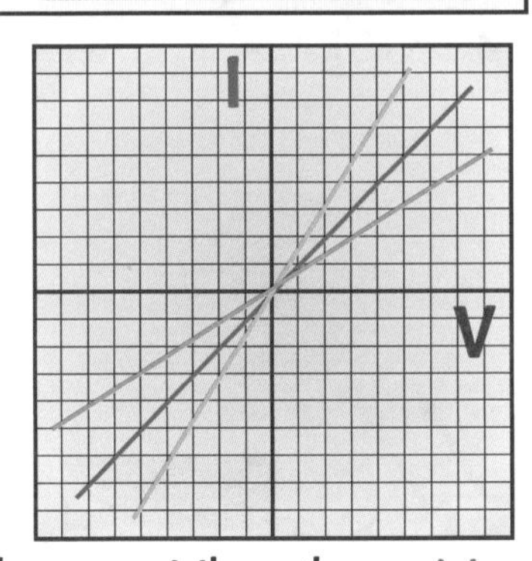

The current through a resistor (at constant temperature) is directly proportional to P.D. Different resistors have different resistances, hence the different slopes.

Filament Lamp

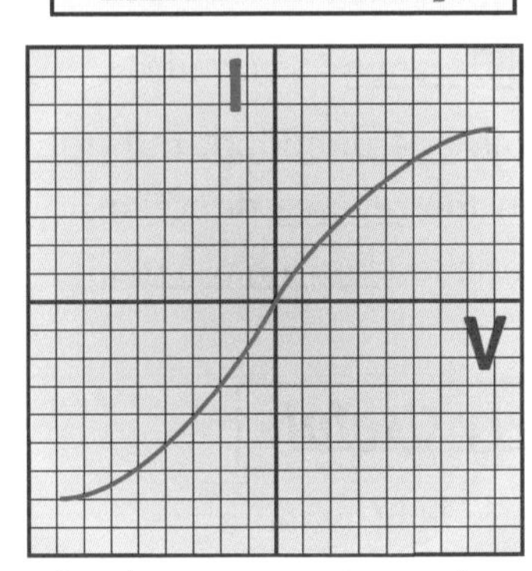

As the temperature of the filament increases, the resistance increases, hence the curve.

Diode

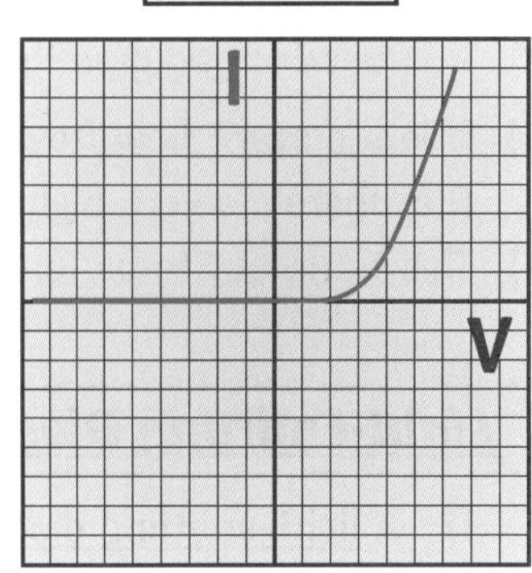

Current will only flow through a diode in one direction, as shown. The diode has very high resistance in the opposite direction.

Resistance Increases with Temperature

1) When an electrical charge flows through a resistor, some of the electrical energy is transferred to heat energy and the resistor gets hot.
2) This heat energy causes the ions in the conductor to vibrate more. With the ions jiggling around it's more difficult for the charge-carrying electrons to get through the resistor — the current can't flow as easily and the resistance increases.
3) For most resistors there is a limit to the amount of current that can flow. More current means an increase in temperature, which means an increase in resistance, which means the current decreases again.
4) This is why the graph for the filament lamp levels off at high currents.

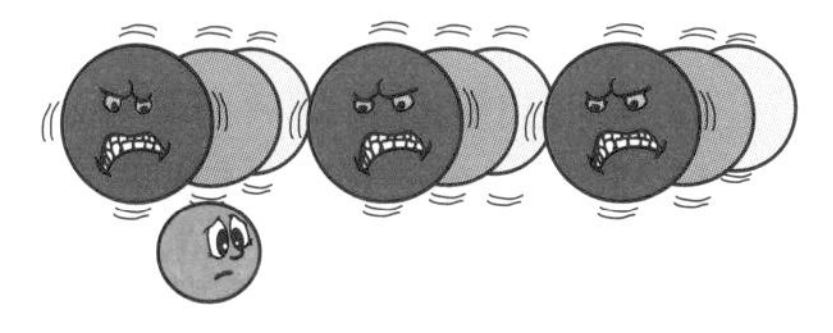

Resistance, Potential Difference and Current: $V = I \times R$

Potential Difference = Current × Resistance

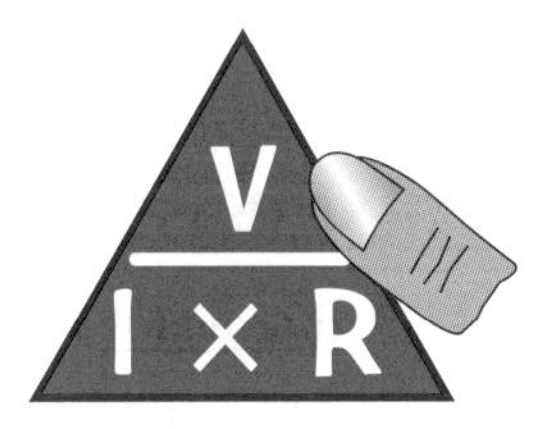

For the straight-line graphs above, the resistance of the component is steady and is equal to the inverse of the gradient of the line, or "1/gradient". In other words, the steeper the graph the lower the resistance.

If the graph curves, it means the resistance is changing. In that case R can be found for any point by taking the pair of values (V, I) from the graph and sticking them in the formula $R = V/I$. Easy.

EXAMPLE: Voltmeter V reads 6 V and resistor R is 4 Ω. What is the current through Ammeter A?

ANSWER: Use the formula triangle for $V = I \times R$. We need to find I, so the version we need is $I = V/R$. The answer is then: $I = 6 \div 4 = 1.5$ A.

In the end you'll have to learn this — resistance is futile...

You have to be able to interpret potential difference-current graphs for your exam. Remember — the steeper the slope, the lower the resistance. And you need to know that formula inside out, back to front, upside down and in Swahili. It's the most important equation in electrics, bar none. (P.S. I might let you off the Swahili.)

Circuit Devices

You might consider yourself a bit of an expert in circuit components — you're enlightened about bulbs, you're switched on to switches... Just make sure you know these ones as well — they're a little bit trickier.

Current Only Flows in One Direction through a Diode

1) A diode is a special device made from semiconductor material such as silicon.
2) It is used to regulate the potential difference in circuits.
3) It lets current flow freely through it in one direction, but not in the other (i.e. there's a very high resistance in the reverse direction).
4) This turns out to be real useful in various electronic circuits.

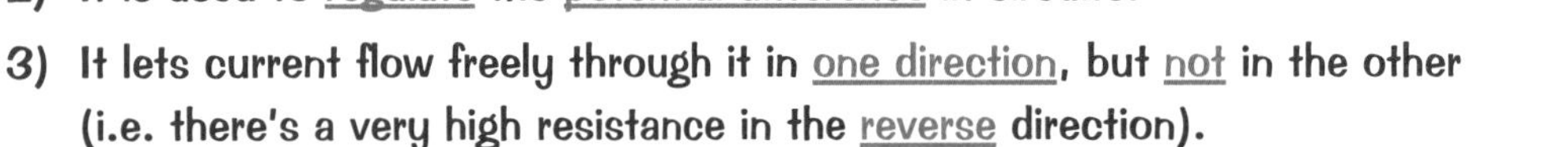

Light-Emitting Diodes are Very Useful

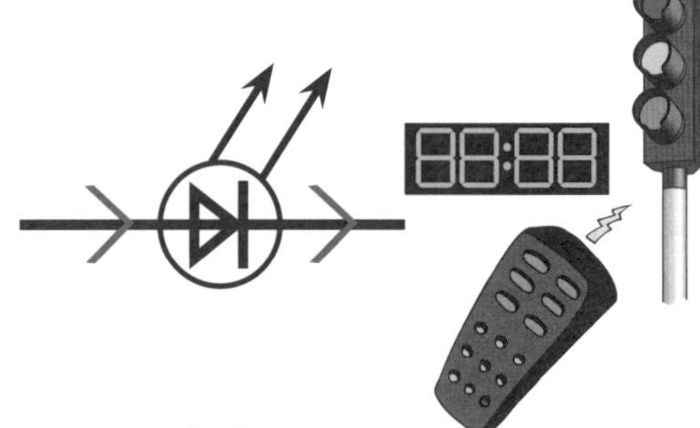

1) A light-emitting diode (LED) emits light when a current flows through it in the forward direction.
2) LEDs are being used more and more as lighting, as they use a much smaller current than other forms of lighting.
3) LEDs indicate the presence of current in a circuit. They're often used in appliances (e.g. TVs) to show that they are switched on.
4) They're also used for the numbers on digital clocks, in traffic lights and in remote controls.

A Light-Dependent Resistor or "LDR" to You

1) An LDR is a resistor that is dependent on the intensity of light. Simple really.
2) In bright light, the resistance falls.
3) In darkness, the resistance is highest.
4) They have lots of applications including automatic night lights, outdoor lighting and burglar detectors.

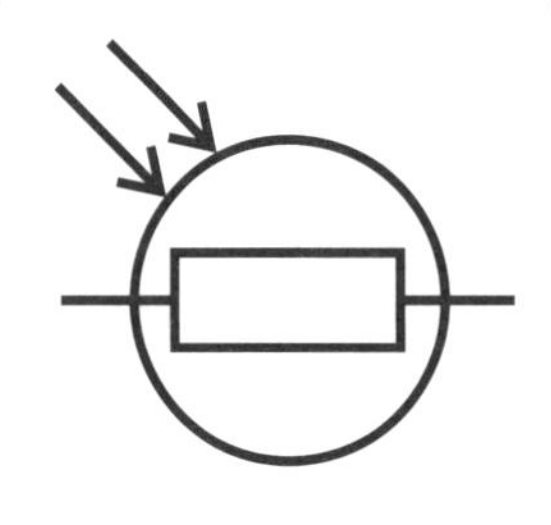

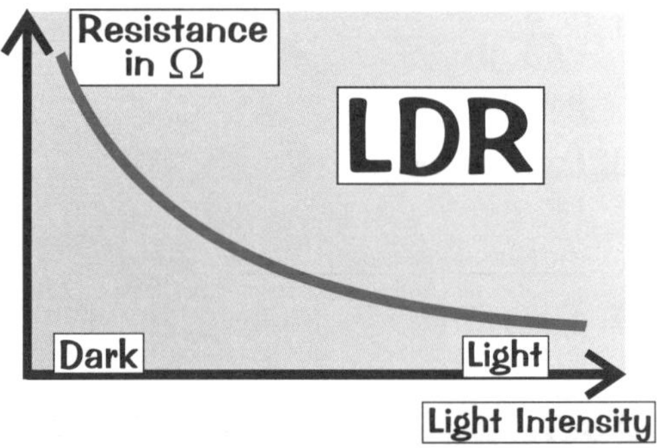

The Resistance of a Thermistor Decreases as Temperature Increases

1) A thermistor is a temperature dependent resistor.
2) In hot conditions, the resistance drops.
3) In cool conditions, the resistance goes up.
4) Thermistors make useful temperature detectors, e.g. car engine temperature sensors and electronic thermostats.

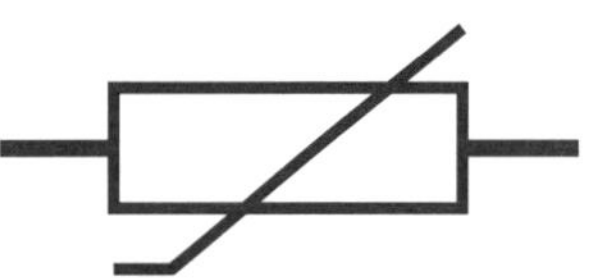

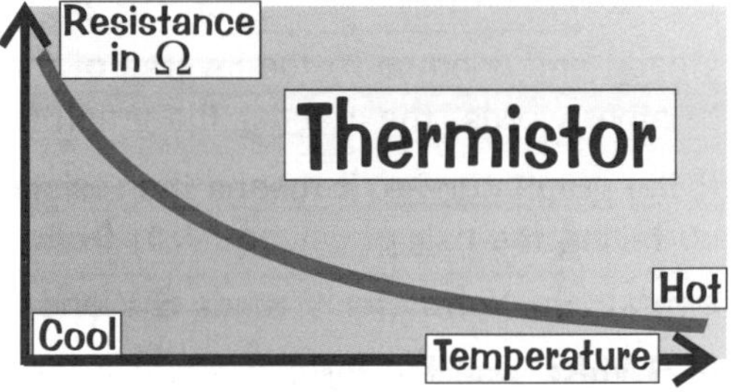

LDRs — Light-Dependent Rabbits...

LDRs are good triggers in security systems, because they can detect when the light intensity changes. So if a robber walks in front of a beam of light pointed at the LDR, the resistance shoots up and an alarm goes off.

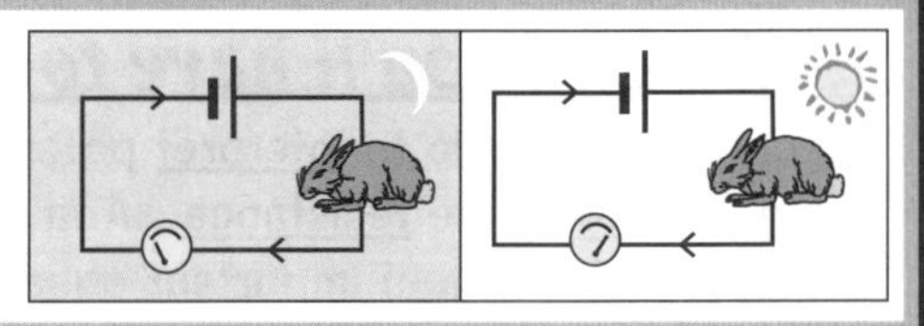

Series Circuits

You need to be able to tell the difference between series and parallel circuits just by looking at them. You also need to know the rules about what happens with both types. Read on.

Series Circuits — All or Nothing

1) In series circuits, the different components are connected in a line, end to end, between the +ve and –ve of the power supply (except for voltmeters, which are always connected in parallel, but they don't count as part of the circuit).
2) If you remove or disconnect one component, the circuit is broken and they all stop.
3) This is generally not very handy, and in practice very few things are connected in series.

1) Potential Difference is Shared:

In series circuits the total P.D. of the supply is shared between the various components. So the voltages round a series circuit always add up to equal the source voltage:

$$V = V_1 + V_2 + \dots$$

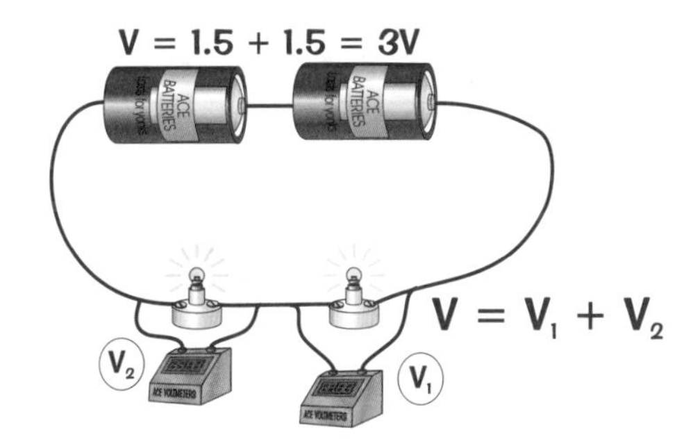

2) Current is the Same Everywhere:

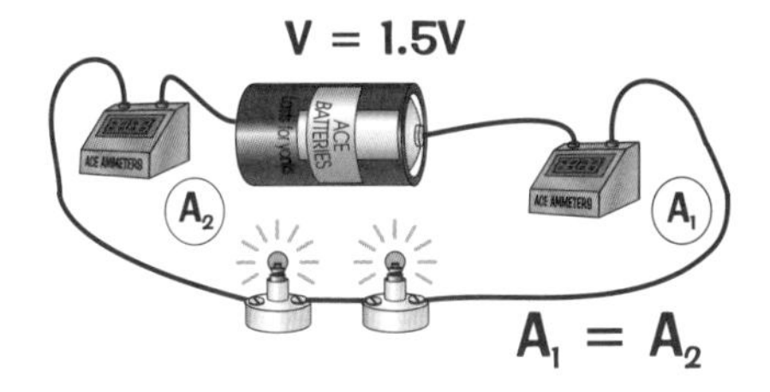

1) In series circuits the same current flows through all parts of the circuit, i.e: $A_1 = A_2$
2) The size of the current is determined by the total P.D. of the cells and the total resistance of the circuit: i.e. $I = V/R$

3) Resistance Adds Up:

1) In series circuits the total resistance is just the sum of all the resistances:

$$R = R_1 + R_2 + R_3$$

2) The bigger the resistance of a component, the bigger its share of the total P.D.

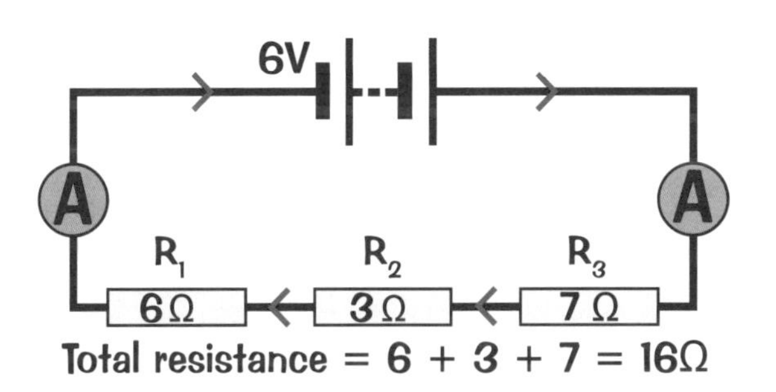

Cell Voltages Add Up:

1) There is a bigger potential difference when more cells are in series, provided the cells are all connected the same way.
2) For example when two batteries of voltage 1.5 V are connected in series they supply 3 V between them.

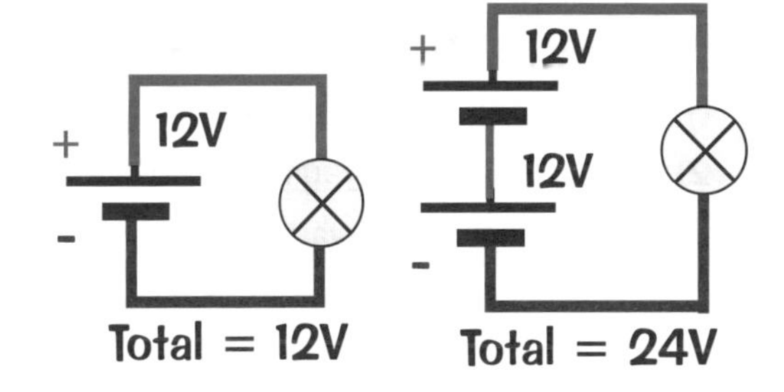

Series circuits — they're no laughing matter...

If you connect a lamp to a battery, it lights up with a certain brightness. If you then add more identical lamps in series with the first one, they'll all light up less brightly than before. That's because in a series circuit the voltage is shared out between all the components. That doesn't happen in parallel circuits...

Parallel Circuits

Parallel circuits are much more sensible than series circuits and so they're much more common in real life. All the electrics in your house will be wired in parallel circuits.

Parallel Circuits — Independence and Isolation

1) In parallel circuits, each component is separately connected to the +ve and –ve of the supply.
2) If you remove or disconnect one of them, it will hardly affect the others at all.
3) This is obviously how most things must be connected, for example in cars and in household electrics. You have to be able to switch everything on and off separately.

1) P.D. is the Same Across All Components:

1) In parallel circuits all components get the full source P.D., so the voltage is the same across all components:

$$V_1 = V_2 = V_3$$

2) This means that identical bulbs connected in parallel will all be at the same brightness.

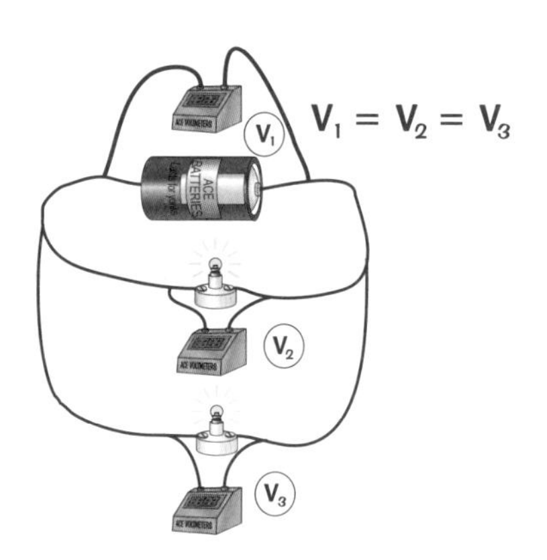

2) Current is Shared Between Branches:

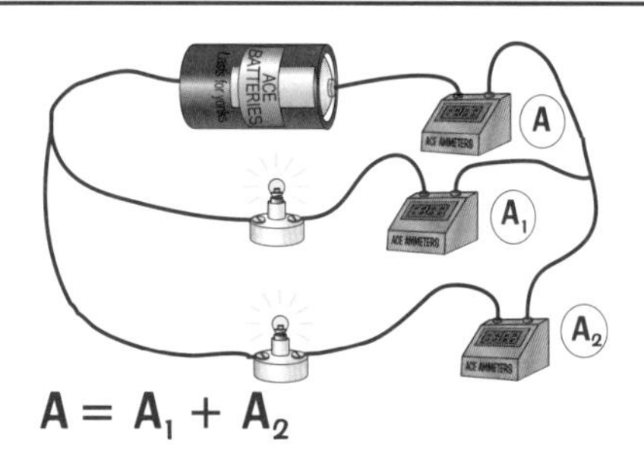

$A = A_1 + A_2$

1) In parallel circuits the total current flowing around the circuit is equal to the total of all the currents through the separate components.

$$A = A_1 + A_2 + ...$$

2) In a parallel circuit, there are junctions where the current either splits or rejoins. The total current going into a junction has to equal the total current leaving.
3) If two identical components are connected in parallel then the same current will flow through each component.

Voltmeters and Ammeters Are Exceptions to the Rule:

1) Ammeters and voltmeters are exceptions to the series and parallel rules.
2) Ammeters are always connected in series even in a parallel circuit.
3) Voltmeters are always connected in parallel with a component even in a series circuit.

A current shared — is a current halved...

Parallel circuits might look a bit scarier than series ones, but they're much more useful — and you don't have to learn as many equations for them (yay!). Remember: each branch has the same voltage across it, and the total current is equal to the sum of the currents through each of the branches.

Series and Parallel Circuits — Examples

It's not enough to know how circuits work in theory, it's important that you can calculate the currents, potential differences and resistances in a range of examples. It will be on the exam, so work through these examples now.

Example on Series Circuits

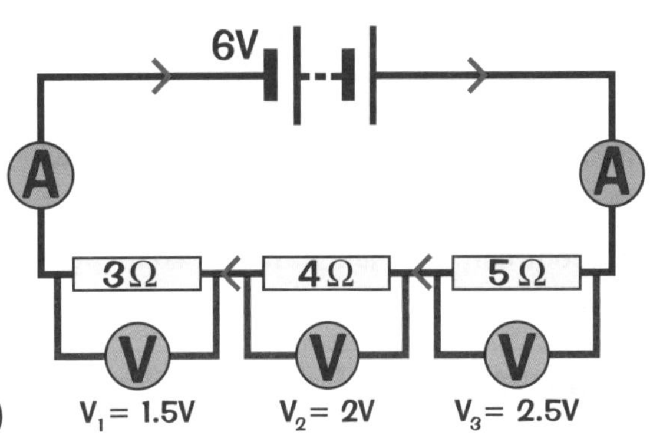

Potential differences add to equal the source P.D.:
1.5 + 2 + 2.5 = 6 V

Total resistance is the sum of the resistances in the circuit:
3 + 4 + 5 = 12 Ω

Current flowing through all parts of the circuit
= V/R = 6/12 = 0.5 A

(If an extra cell was added of P.D. 3 V then the P.D. across each resistor would increase and the current would increase too.)

Christmas Fairy Lights are Often Wired in Series

Christmas fairy lights are about the only real-life example of things connected in series, and we all know what a pain they are when the whole lot go out just because one of the bulbs is slightly dicky. The only advantage is that the bulbs can be very small because the total 230 V is shared out between them, so each bulb only has a small potential difference across it.

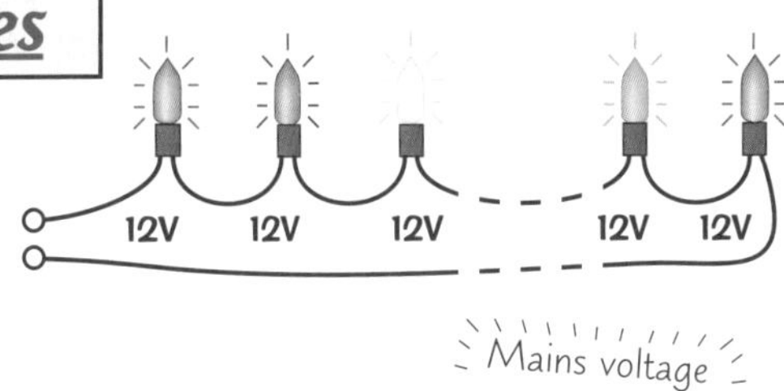

Example on Parallel Circuits

1) The P.D. across each resistor in the circuit is the same as the supply P.D. Each voltmeter will read 6 V.
2) The current through each resistor will be different because they have different values of resistance.
3) The current through the battery is the same as the sum of the other currents in the branches.
   i.e. $A_1 = A_2 + A_3 + A_4 \Rightarrow A_1 = 1.5 + 3 + 1 = 5.5$ A

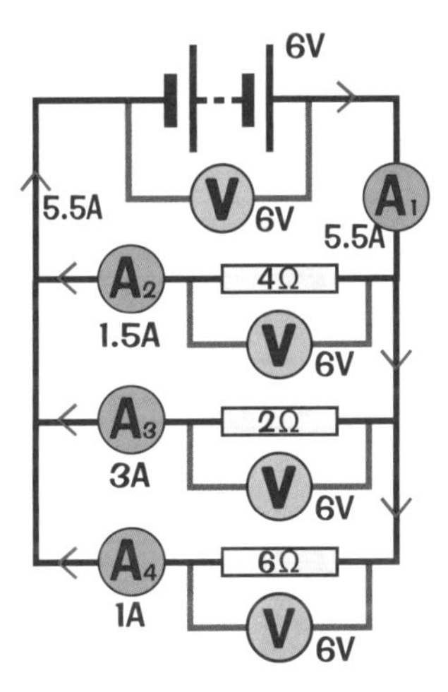

Everything Electrical in a Car is Connected in Parallel

Parallel connection is essential in a car to give these two features:

1) Everything can be turned on and off separately.
2) Everything always gets the full voltage from the battery.

The only slight effect is that when you turn lots of things on the lights may go dim because the battery can't provide full voltage under heavy load. This is normally a very slight effect. You can spot the same thing at home when you turn a kettle on, if you watch very carefully.

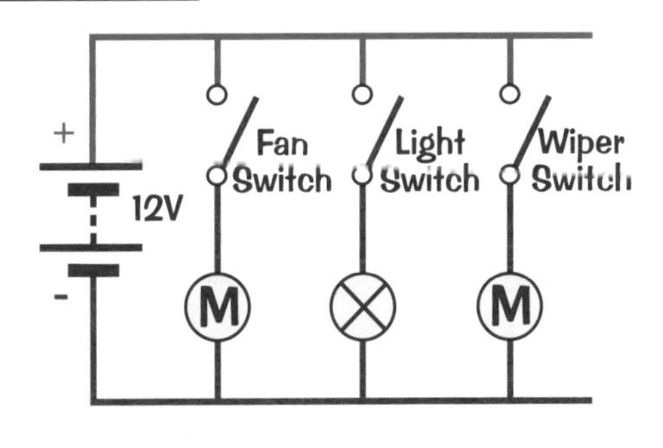

Ⓜ is the symbol for a motor.

In a parallel universe — my car would start...

A lot of fairy lights are actually done on a parallel circuit these days — they have an adapter that brings the voltage down, so the lights can still be diddy but it doesn't matter if one of them blows. Cunning.

Mains Electricity

Electric current is the movement of charge carriers. To transfer energy, it doesn't matter which way the charge carriers are going. That's why an alternating current works. Read on to find out more...

Mains Supply is AC, Battery Supply is DC

1) The UK mains supply is approximately 230 volts.
2) It is an AC supply (alternating current), which means the current is constantly changing direction.
3) The frequency of the AC mains supply is 50 cycles per second or 50 Hz (hertz).
4) By contrast, cells and batteries supply direct current (DC). This just means that the current always keeps flowing in the same direction.

Electricity Supplies Can Be Shown on an Oscilloscope Screen

1) A cathode ray oscilloscope (CRO) is basically a snazzy voltmeter.
2) If you plug an AC supply into an oscilloscope, you get a 'trace' on the screen that shows how the voltage of the supply changes with time. The trace goes up and down in a regular pattern — some of the time it's positive and some of the time it's negative.
3) If you plug in a DC supply, the trace you get is just a straight line.

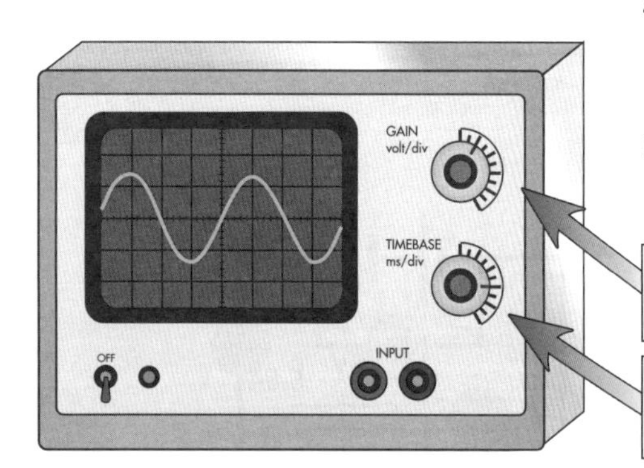

4) The vertical height of the AC trace at any point shows the input voltage at that point. By measuring the height of the trace you can find the potential difference of the AC supply.
5) For DC it's a lot simpler — the voltage is just the distance from the straight line trace to the centre line.

The GAIN dial controls how many volts each centimetre division represents on the vertical axis.

The TIMEBASE dial controls how many milliseconds (1 ms = 0.001 s) each division represents on the horizontal axis.

Learn How to Read an Oscilloscope Trace

DC supply

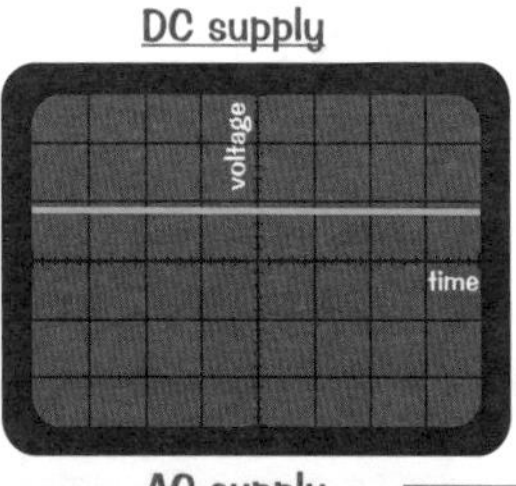

A DC source is always at the same voltage, so you get a straight line.

AC supply

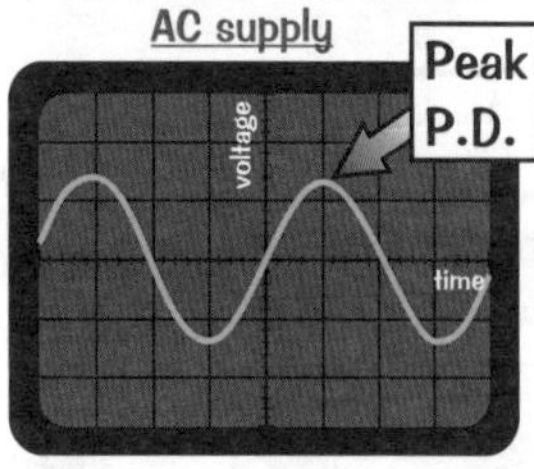

An AC source gives a regularly repeating wave. From that, you can work out the period and the frequency of the supply.

You work out the frequency using:

$$\text{Frequency (Hz)} = \frac{1}{\text{Time period (s)}}$$

EXAMPLE: The trace below comes from an oscilloscope with the timebase set to 5 ms/div. Find: a) the time period, and b) the frequency of the AC supply.

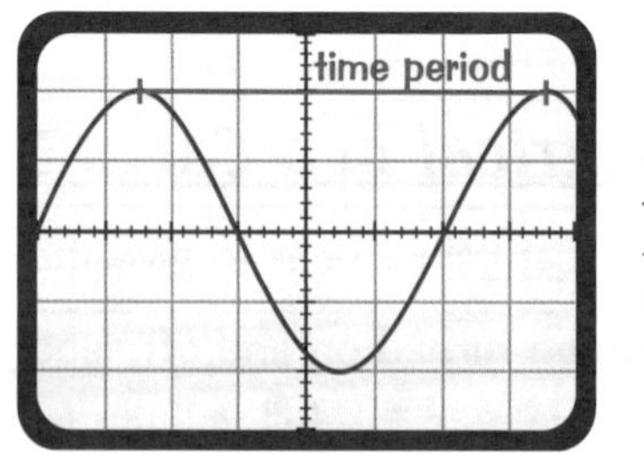

Time period = the time to complete one cycle. 1 ms = 0.001 s.

ANSWER: a) To find the time period, measure the horizontal distance between two peaks. The time period of the signal is 6 divisions. Multiply by the timebase:
Time period = 5 ms × 6 = 0.03 s

b) Using the frequency formula on the left:
Frequency = 1/0.03 = 33 Hz

I wish my bank account had a gain dial...

Because mains power is AC, its current can be increased or decreased using a device called a transformer. The lower the current in power transmission lines, the less energy is wasted as heat.

Electricity in the Home

Now then, did you know... electricity is dangerous. It can kill you. Well just watch out for it, that's all.

Hazards in the Home — Eliminate Them Before They Eliminate You

A likely exam question will show you a picture of domestic bliss but with various electrical hazards in the picture such as kids shoving their fingers into sockets and stuff like that, and they'll ask you to list all the hazards. This should be mostly common sense, but it won't half help if you already know some of the likely hazards, so learn these 9 examples:

1) Long cables.
2) Frayed cables.
3) Cables in contact with something hot or wet.
4) Water near sockets.
5) Shoving things into sockets.
6) Damaged plugs.
7) Too many plugs into one socket.
8) Lighting sockets without bulbs in.
9) Appliances without their covers on.

Most Cables Have Three Separate Wires

1) Most electrical appliances are connected to the mains supply by three-core cables. This means that they have three wires inside them, each with a core of copper and a coloured plastic coating.
2) The brown LIVE WIRE in a mains supply alternates between a HIGH +VE AND –VE VOLTAGE.
3) The blue NEUTRAL WIRE is always at 0V. Electricity normally flows in and out through the live and neutral wires only.
4) The green and yellow EARTH WIRE is for protecting the wiring, and for safety — it works together with a fuse to prevent fire and shocks. It is attached to the metal casing of the plug and carries the electricity to earth (and away from you) should something go wrong and the live or neutral wires touch the metal case.

live wire (alternating between +ve and –ve high voltage)

neutral wire (0V)

earth wire

insulating sheath

Three-Pin Plugs and Cables — Learn the Safety Features

Get the Wiring Right

1) The right coloured wire is connected to each pin, and firmly screwed in.
2) No bare wires showing inside the plug.
3) Cable grip tightly fastened over the cable outer layer.
4) Different appliances need different amounts of electrical energy. Thicker cables have less resistance, so they carry more current.

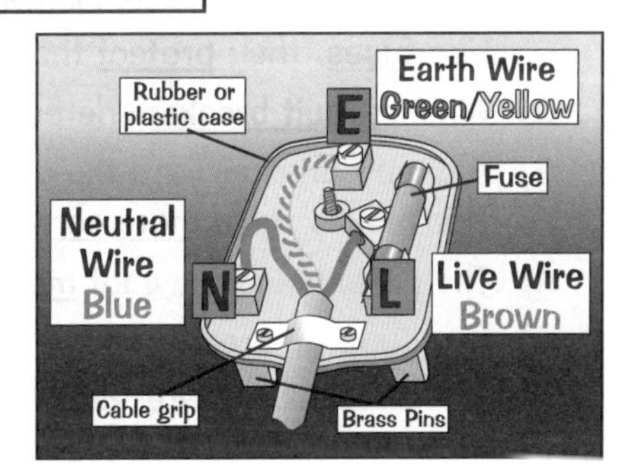

Plug Features

1) The metal parts are made of copper or brass because these are very good conductors.
2) The case, cable grip and cable insulation are made of rubber or plastic because they're really good insulators, and flexible too.
3) This all keeps the electricity flowing where it should.

CGP books are ACE — well, I had to get a plug in somewhere...

Pure water doesn't conduct electricity, but water (usually) has mineral salts dissolved in it. These carry the charge around really well, making it a very good conductor. So don't blow dry your hair in the bath, OK?

Fuses and Earthing

Questions about fuses are an exam favourite because they cover a whole barrel of fun — electrical current, resistance, energy transfers and electrical safety. Learn this page and make sure you've got it sussed.

Earthing and Fuses Prevent Electrical Overloads

The earth wire and fuse (or circuit breaker) are included in electrical appliances for safety and work together like this:

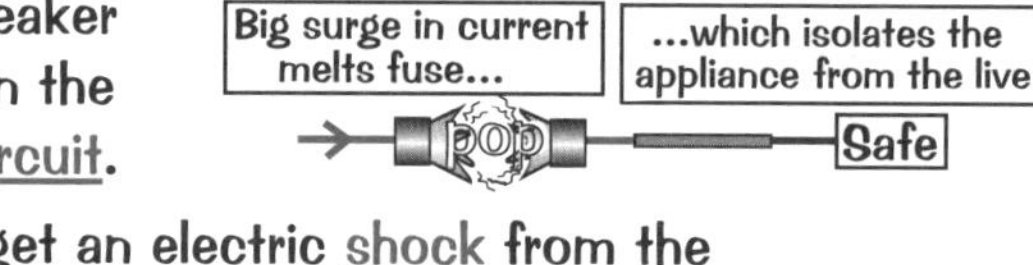

1) If a fault develops in which the live wire somehow touches the metal case, then because the case is earthed, too great a current flows in through the live wire, through the case and out down the earth wire.
2) This surge in current melts the fuse (or trips the circuit breaker in the live wire) when the amount of current is greater than the fuse rating. This cuts off the live supply and breaks the circuit.
3) This isolates the whole appliance, making it impossible to get an electric shock from the case. It also prevents the risk of fire caused by the heating effect of a large current.
4) As well as people, fuses and earthing are there to protect the circuits and wiring in your appliances from getting fried if there is a current surge.
5) Fuses should be rated as near as possible but just higher than the normal operating current.
6) The larger the current, the thicker the cable you need to carry it. That's why the fuse rating needed for cables usually increases with cable thickness.

Insulating Materials Make Appliances "Double Insulated"

All appliances with metal cases are usually "earthed" to reduce the danger of electric shock. "Earthing" just means the case must be attached to an earth wire. An earthed conductor can never become live. If the appliance has a plastic casing and no metal parts showing then it's said to be double insulated.

Anything with double insulation like that doesn't need an earth wire — just a live and neutral. Cables that only carry the live and neutral wires are known as two-core cables.

Circuit Breakers Have Some Advantages Over Fuses

1) Circuit breakers are an electrical safety device used in some circuits. Like fuses, they protect the circuit from damage if too much current flows.
2) When circuit breakers detect a surge in current in a circuit, they break the circuit by opening a switch.
3) A circuit breaker (and the circuit they're in) can easily be reset by flicking a switch on the device. This makes them more convenient than fuses — which have to be replaced once they've melted.
4) They are, however, a lot more expensive to buy than fuses.
5) One type of circuit breaker used instead of a fuse and an earth wire is a Residual Current Circuit Breakers (RCCBs):

    a) Normally exactly the same current flows through the live and neutral wires. If somebody touches the live wire, a small but deadly current will flow through them to the earth. This means the neutral wire carries less current than the live wire. The RCCB detects this difference in current and quickly cuts off the power by opening a switch.

    b) They also operate much faster than fuses — they break the circuit as soon as there is a current surge — no time is wasted waiting for the current to melt a fuse. This makes them safer.

    c) RCCBs even work for small current changes that might not be large enough to melt a fuse. Since even small current changes could be fatal, this means RCCBs are more effective at protecting against electrocution.

Why are earth wires green and yellow — when mud is brown..?

All these safety precautions mean it's pretty difficult to get electrocuted on modern appliances. But that's only so long as they are in good condition and you're not doing something really stupid. Watch out for frayed wires, don't overload plugs, and for goodness sake don't use a knife to get toast out of a toaster when it is switched on.

Energy and Power in Circuits

Electricity is just another form of energy — which means that it is always conserved.

Energy is Transferred from Cells and Other Sources

Anything which supplies electricity is also supplying energy.
So cells, batteries, generators, etc. all transfer energy to components in the circuit:

Motion: motors | Light: light bulbs | Heat: Hair dryers/kettles | Sound: speakers

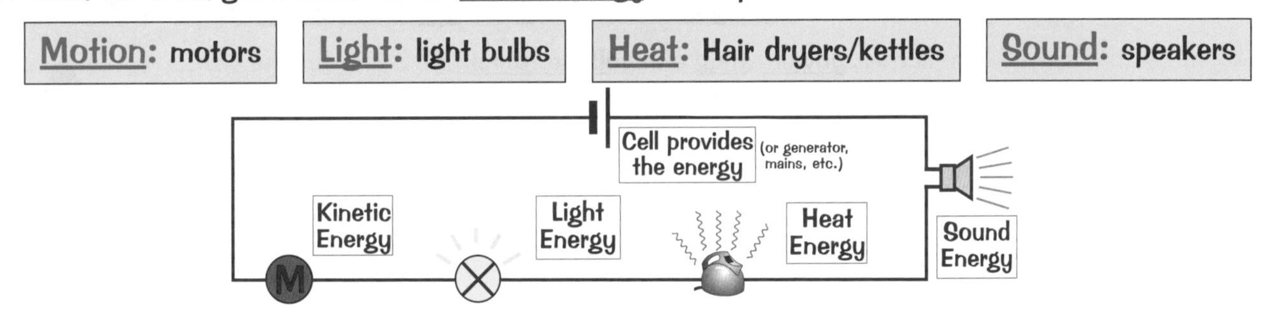

All Resistors Produce Heat When a Current Flows Through Them

1) Whenever a current flows through anything with electrical resistance (which is pretty much everything) then electrical energy is converted into heat energy.
2) The more current that flows, the more heat is produced.
3) A bigger voltage means more heating because it pushes more current through.
4) Filament bulbs work by passing a current through a very thin wire, heating it up so much that it glows. Rather obviously, they waste a lot of energy as heat.

If an Appliance is Efficient it Wastes Less Energy

All this energy wasted as heat can get a little depressing — but there is a solution.

1) When you buy electrical appliances you can choose to buy ones that are more energy efficient.
2) These appliances transfer more of their total electrical energy output to useful energy.
3) For example, less energy is wasted as heat in power-saving lamps such as compact fluorescent lamps (CFLs) and light emitting diodes (p.62) than in ordinary filament bulbs.
4) Unfortunately, they do cost more to buy, but over time the money you save on your electricity bills pays you back for the initial investment.

Not an energy efficient lamp.

Power Ratings of Appliances

The total energy transferred by an appliance depends on how long the appliance is on and its power rating.
The power of an appliance is the energy that it uses per second.

Energy Transferred = Power rating × time

$$\frac{E}{P \times t}$$

For example, if a 2.5 kW kettle is on for 5 minutes, the energy transferred by the kettle in this time is $300 \times 2500 = 750\ 000$ J = 750 kJ. (5 minutes = 300 s).

Ohm's girlfriend was a vixen — he couldn't resistor...

The equation for power is a real simple one, but it's absolutely essential that you've got it hard-wired into your memory. Remember: power is energy transferred per second. Power is energy transferred per second. Power is energy transferred per second....

Power and Energy Change

You can think about electrical circuits in terms of energy transfer — the charge carriers take charge around the circuit, and when they go through an electrical component energy is transferred to make the component work.

Electrical Power and Fuse Ratings

1) The formula for electrical power is: **POWER = CURRENT × POTENTIAL DIFFERENCE** $P = I \times V$

2) Most electrical goods show their power rating and voltage rating. To work out the size of the fuse needed, you need to work out the current that the item will normally use:

EXAMPLE: A hair dryer is rated at 230 V, 1 kW. Find the fuse needed.
ANSWER: I = P/V = 1000/230 = 4.3 A. Normally, the fuse should be rated just a little higher than the normal current, so a 5 amp fuse is ideal for this one.

The Potential Difference is the Energy Transferred per Charge Passed

1) When an electrical charge (Q) goes through a change in potential difference (V), then energy (E) is transferred.
2) Energy is supplied to the charge at the power source to 'raise' it through a potential.
3) The charge gives up this energy when it 'falls' through any potential drop in components elsewhere in the circuit.

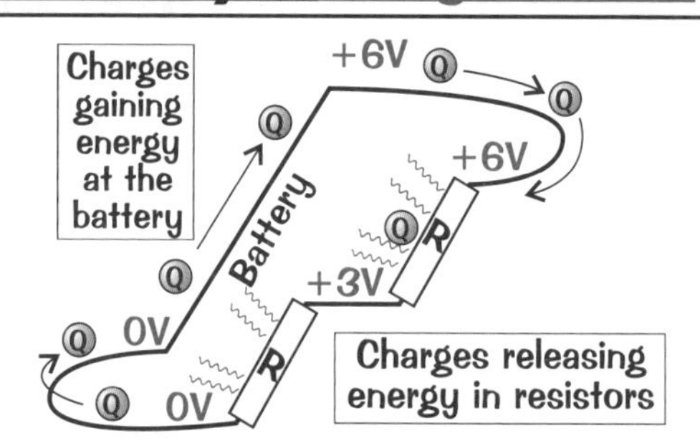

The formula is real simple:

Energy transformed = Charge × Potential difference

4) The bigger the change in P.D. (or voltage.), the more energy is transferred for a given amount of charge passing through the circuit.
5) That means that a battery with a bigger voltage will supply more energy to the circuit for every coulomb of charge which flows round it, because the charge is raised up "higher" at the start (see above diagram) — and as the diagram shows, more energy will be dissipated in the circuit too.

EXAMPLE: The motor in an electric toothbrush is attached to a 3 V battery. If a current of 0.8 A flows through the motor for 3 minutes:
a) Calculate the total charge passed.
b) Calculate the energy transformed by the motor.
c) Explain why the kinetic energy output of the motor will be less than your answer to b).

ANSWER: a) Use the formula (p.59) Q = I × t = 0.8 × (3 × 60) = 144 C

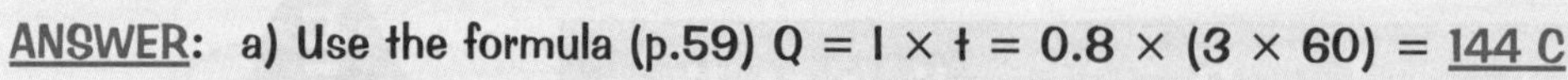

b) Use E = Q × V = 144 × 3 = 432 J
c) The motor won't be 100% efficient. Some of the energy will be transformed into sound and heat.

You have the power — now use your potential...

Ok, another two formulas to learn. By this point you're probably experiencing a little bit of formula fatigue, but trust me, you will be glad that you learned them all. Try to think about exactly what each one means and how they work together — things are a lot easier to memorise if you have a real understanding of why they are there.

Atomic Structure

Ernest Rutherford didn't just pick the nuclear model of the atom out of thin air. It all started with a Greek fella called Democritus in the 5th Century BC. He thought that all matter, whatever it was, was made up of identical lumps called "atomos". And that's about as far as the theory got until the 1800s...

Rutherford Scattering and the Demise of the Plum Pudding

1) In 1804 John Dalton agreed with Democritus that matter was made up of tiny spheres ("atoms") that couldn't be broken up, but he reckoned that each element was made up of a different type of "atom".
2) Nearly 100 years later, J J Thomson discovered that electrons could be removed from atoms. So Dalton's theory wasn't quite right (atoms could be broken up). Thomson suggested that atoms were spheres of positive charge with tiny negative electrons stuck in them like plums in a plum pudding.
3) That "plum pudding" theory didn't last very long though. In 1909 Rutherford and Marsden tried firing a beam of alpha particles (see p.73) at thin gold foil. They expected that the positively charged alpha particles would be slightly deflected by the electrons in the plum pudding model.
4) However, most of the alpha particles just went straight through, but the odd one came straight back at them, which was frankly a bit of a shocker for Rutherford and his pal.

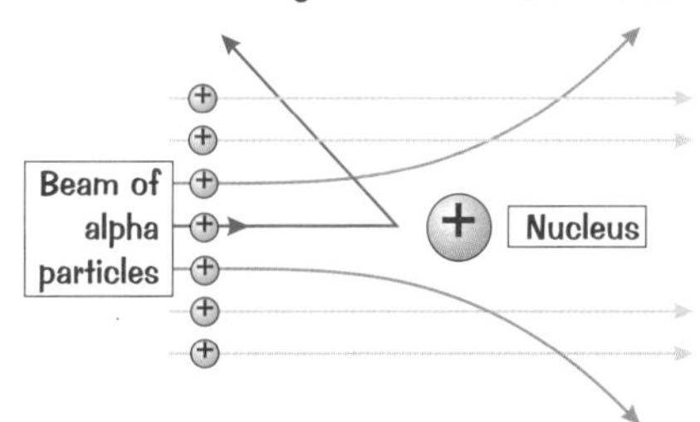

5) Being pretty clued-up guys, Rutherford and Marsden realised this meant that most of the mass of the atom was concentrated at the centre in a tiny nucleus. They also realised that the nucleus must have a positive charge, since it repelled the positive alpha particles.
6) It also showed that most of an atom is just empty space, which is also a bit of a shocker when you think about it.

Rutherford and Marsden Came Up with the Nuclear Model of the Atom

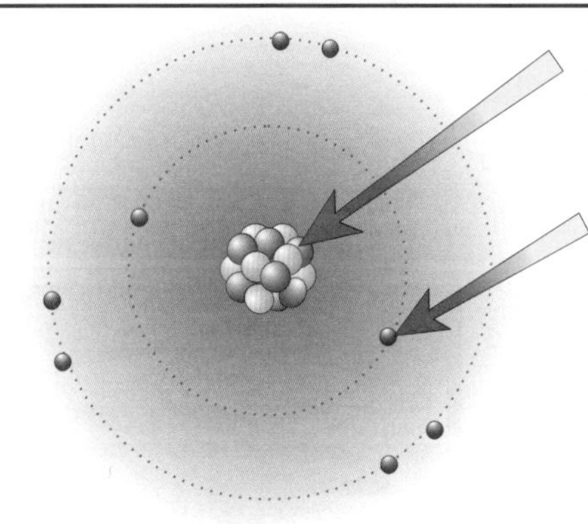

The nucleus is tiny but it makes up most of the mass of the atom. It contains protons (which are positively charged) and neutrons (which are neutral) — which gives it an overall positive charge.

The rest of the atom is mostly empty space. The negative electrons whizz round the outside of the nucleus really fast. They give the atom its overall size — the radius of the atom's nucleus is about 10 000 times smaller than the radius of the atom. Crikey.

Learn the relative charges and masses of each particle:

PARTICLE	MASS	CHARGE
Proton	1	+1
Neutron	1	0
Electron	1/2000	- 1

Number of Protons Equals Number of Electrons

1) Atoms have no charge overall.
2) The charge on an electron is the same size as the charge on a proton — but opposite.
3) This means the number of protons always equals the number of electrons in a neutral atom.
4) If some electrons are added or removed, the atom becomes a charged particle called an ion.

And I always thought Kate Moss was the best model...

The nuclear model is just one way of thinking about the atom. It works really well for explaining a lot of physical properties of different elements, but it's certainly not the whole story. Other bits of science are explained using different models of the atom. The beauty of it though is that no one model is more right than the others.

Atoms and Ionising Radiation

You have just entered the subatomic realm — now stuff starts to get real interesting...

Isotopes are Different Forms of the Same Element

1) Isotopes are atoms with the same number of protons but a different number of neutrons.
2) Hence they have the same atomic number, but different mass numbers.
3) Atomic number is the number of protons in an atom.
4) Mass number is the number of protons + the number of neutrons in an atom.
5) Carbon-12 and carbon-14 are good examples of isotopes:

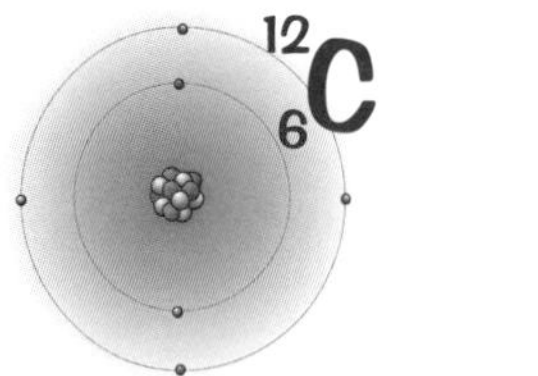

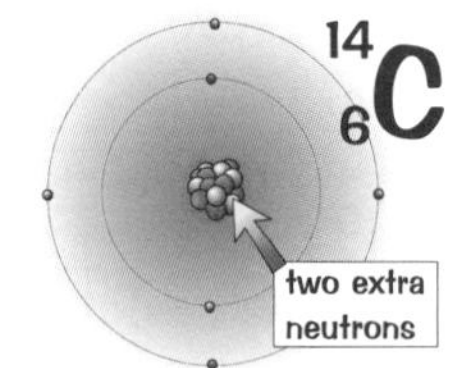

6) Most elements have different isotopes, but there's usually only one or two stable ones.
7) The other isotopes tend to be radioactive, which means they decay into other elements and give out radiation.

Radioactivity is a Totally Random Process

1) Radioactive substances give out radiation from the nuclei of their atoms — no matter what is done to them.
2) This process is entirely random. This means that if you have 1000 unstable nuclei, you can't say when any one of them is going to decay, and neither can you do anything at all to make a decay happen.
3) It's completely unaffected by physical conditions like temperature or by any sort of chemical bonding etc.
4) Radioactive substances spit out one or more of the three types of radiation, alpha, beta or gamma (see next page).

Background Radiation Comes from Many Sources

Background radiation is radiation that is present at all times, all around us, wherever you go. The background radiation we receive comes from:

1) Radioactivity of naturally occurring unstable isotopes which are all around us — in the air, in food, in building materials and in the rocks under our feet.

2) Radiation from space, which is known as cosmic rays. These come mostly from the Sun.

3) Radiation due to man-made sources, e.g. fallout from nuclear weapons tests, nuclear accidents (such as Chernobyl) or dumped nuclear waste.

The RELATIVE PROPORTIONS of background radiation:

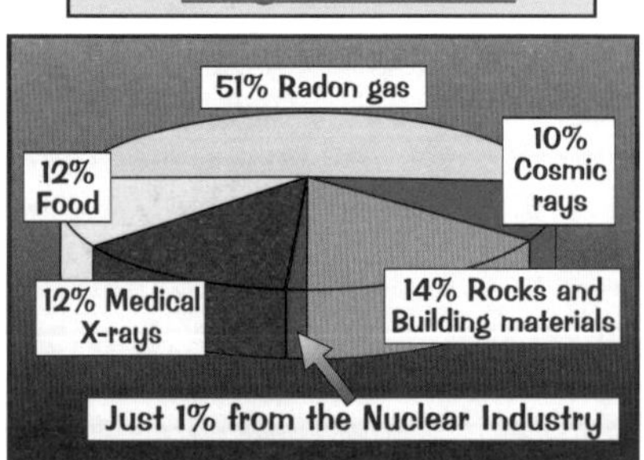

Completely random — just like your revision shouldn't be...

It's the number of protons which decides what element something is, then the number of neutrons decides what isotope of that element it is. And it's unstable isotopes which undergo radioactive decay.

Atoms and Ionising Radiation

Alpha (α) Beta (β) Gamma (γ) — there's a short alphabet of radiation for you to learn here. And it's all ionising.

Alpha Particles are Helium Nuclei

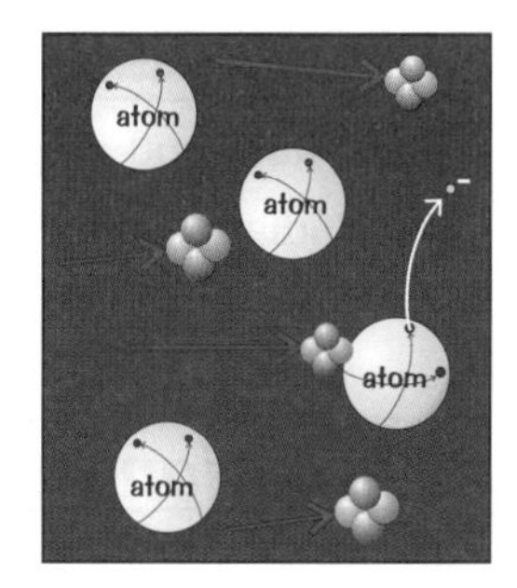

1) An alpha particle is two neutrons and two protons — the same as a helium nucleus.
2) They are relatively big and heavy and slow moving.
3) They therefore don't penetrate very far into materials and are stopped quickly, even when travelling through air.
4) Because of their size they are strongly ionising, which just means they bash into a lot of atoms and knock electrons off them before they slow down, which creates lots of ions — hence the term "ionising".

Beta Particles are Electrons

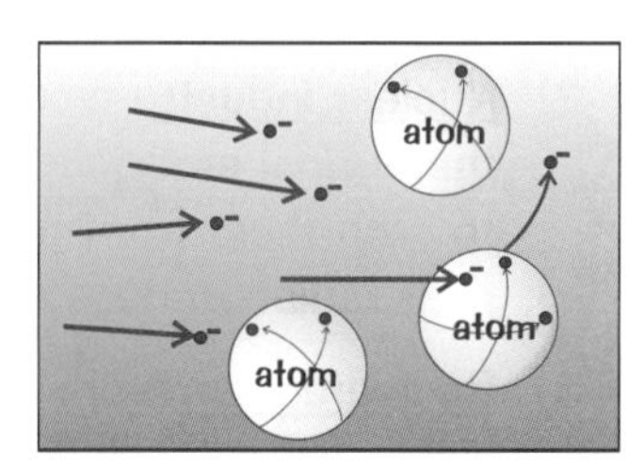

1) Beta particles are in between alpha and gamma in terms of their properties.
2) They move quite fast and they are quite small (they're electrons).
3) They penetrate moderately into materials before colliding, have a long range in air, and are moderately ionising too.
4) For every β-particle emitted, a neutron turns to a proton in the nucleus.
5) A β-particle is simply an electron, with virtually no mass and a charge of –1.

You Need to be Able to Balance Nuclear Equations

1) You can write alpha and beta decays as nuclear equations.
2) Watch out for the mass and atomic numbers — they have to balance up on both sides.

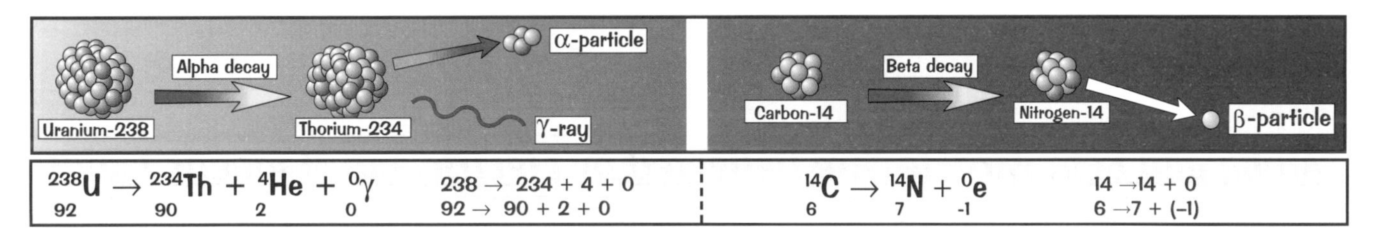

$^{238}_{92}U \rightarrow {}^{234}_{90}Th + {}^{4}_{2}He + {}^{0}_{0}\gamma$ $238 \rightarrow 234 + 4 + 0$, $92 \rightarrow 90 + 2 + 0$

$^{14}_{6}C \rightarrow {}^{14}_{7}N + {}^{0}_{-1}e$ $14 \rightarrow 14 + 0$, $6 \rightarrow 7 + (-1)$

Gamma Rays are Very Short Wavelength EM Waves

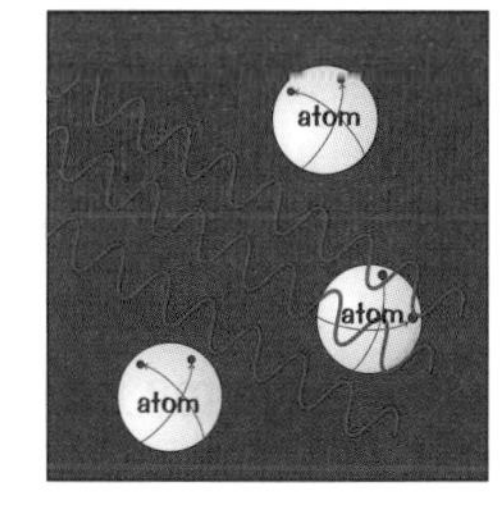

1) Gamma rays are the opposite of alpha particles in a way.
2) They penetrate far into materials without being stopped and pass straight through air.
3) This means they are weakly ionising because they tend to pass through rather than collide with atoms. Eventually they hit something and do damage.
4) Gamma rays have no mass and no charge.

I once beta particle — it cried for ages...

So, when a nucleus decays by alpha emission, its atomic number goes down by two and its mass number goes down by four. Beta emission increases the atomic number by one (the mass number doesn't change).

Atoms and Ionising Radiation

Ooh, it's a mixed bag this page. First up, what affects how much radiation we're exposed to. Then a look at what happens to alpha and beta particles in magnetic and electric fields... And a partridge in a pear treeeeee.

The Damage Caused By Radiation Depends on the Radiation Dose

How likely you are to suffer damage if you're exposed to nuclear radiation depends on the radiation dose.

1) Radiation dose depends on the type and amount of radiation you've been exposed to.
2) The higher the radiation dose, the more at risk you are of developing cancer.

Radiation Dose Depends on Location and Occupation

The amount of radiation you're exposed to (and hence your radiation dose) can be affected by your location and occupation:

Coloured bits indicate more radiation from rocks

1) Certain underground rocks (e.g. granite) can cause higher levels at the surface, especially if they release radioactive radon gas, which tends to get trapped inside people's houses.
2) Nuclear industry workers and uranium miners are typically exposed to 10 times the normal amount of radiation. They wear protective clothing and face masks to stop them from touching or inhaling the radioactive material, and monitor their radiation doses with special radiation badges and regular check-ups.
3) Radiographers work in hospitals using ionising radiation and so have a higher risk of radiation exposure. They wear lead aprons and stand behind lead screens to protect them from prolonged exposure to radiation.

4) At high altitudes (e.g. in jet planes) the background radiation increases because of more exposure to cosmic rays. That means commercial pilots have an increased risk of getting some types of cancer.
5) Underground (e.g. in mines, etc.) it increases because of the rocks all around, posing a risk to miners.

Alpha and Beta Particles are Deflected by Electric and Magnetic Fields

1) Alpha particles have a positive charge, beta particles have a negative charge.
2) When travelling through a magnetic or electric field, both alpha and beta particles will be deflected.
3) They're deflected in opposite directions because of their opposite charge.
4) Alpha particles have a larger charge than beta particles, and feel a greater force in magnetic and electric fields. But they're deflected less because they have a much greater mass.
5) Gamma radiation is an electromagnetic (EM) wave and has no charge, so it doesn't get deflected by electric or magnetic fields.

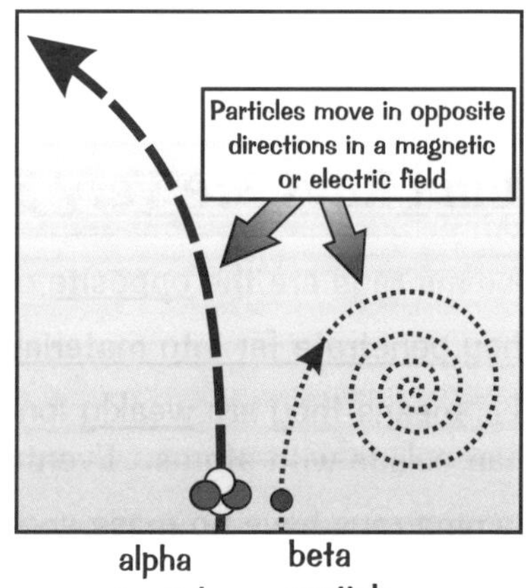

Beta particles in magnetic field? This page has spiralled out of control...

So the amount of radiation you're exposed to depends on your job and your location. Don't forget that some places have higher levels of background radiation than others — so the people there'll get a higher radiation dose.

Half-Life

The unit for measuring radioactivity is the becquerel (Bq). 1 Bq means one nucleus decaying per second.

The Radioactivity of a Sample Always Decreases Over Time

1) This is pretty obvious when you think about it. Each time a decay happens and an alpha, beta or gamma is given out, it means one more radioactive nucleus has disappeared.
2) Obviously, as the unstable nuclei all steadily disappear, the activity (the number of nuclei that decay per second) will decrease. So the older a sample becomes, the less radiation it will emit.

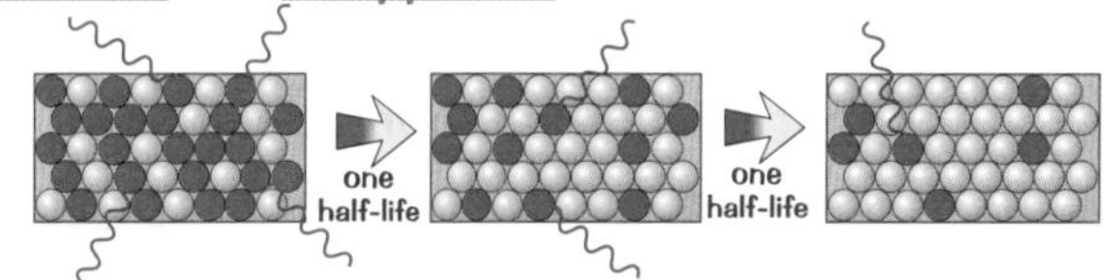

3) How quickly the activity drops off varies a lot. For some substances it takes just a few microseconds before nearly all the unstable nuclei have decayed, whilst for others it can take millions of years.
4) The problem with trying to measure this is that the activity never reaches zero, which is why we have to use the idea of half-life to measure how quickly the activity drops off.
5) Learn this definition of half-life:

HALF-LIFE is the AVERAGE TIME it takes for the NUMBER OF NUCLEI in a RADIOACTIVE ISOTOPE SAMPLE to HALVE.

6) In other words, it is the time it takes for the count rate (the number of radioactive emissions detected per unit of time) from a sample containing the isotope to fall to half its initial level.
7) A short half-life means the activity falls quickly, because lots of the nuclei decay quickly.
8) A long half-life means the activity falls more slowly because most of the nuclei don't decay for a long time — they just sit there, basically unstable, but kind of biding their time.

Do Half-life Questions Step by Step

Half-life is maybe a little confusing, but exam calculations are straightforward so long as you do them slowly, STEP BY STEP. Like this one:

A VERY SIMPLE EXAMPLE: The activity of a radioisotope is 640 cpm (counts per minute). Two hours later it has fallen to 80 cpm. Find the half-life of the sample.

ANSWER: You must go through it in short simple steps like this:

INITIAL count:		after ONE half-life:		after TWO half-lives:		after THREE half-lives:
640	(÷2)→	320	(÷2)→	160	(÷2)→	80

Notice the careful step-by-step method, which tells us it takes three half-lives for the activity to fall from 640 to 80. Hence two hours represents three half-lives, so the half-life is 120 mins ÷ 3 = 40 minutes.

Finding the Half-life of a Sample Using a Graph

1) The data for the graph will usually be several readings of count rate taken with a G-M tube and counter.
2) The graph will always be shaped like the one shown.
3) The half-life is found from the graph by finding the time interval on the bottom axis corresponding to a halving of the activity on the vertical axis. Easy peasy really.

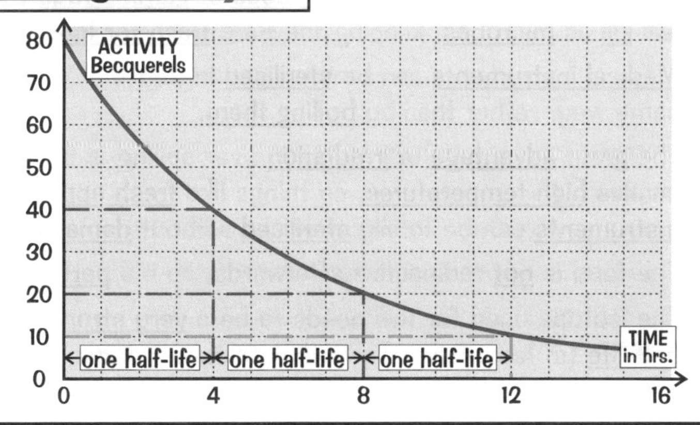

Half-life of a box of chocolates — about five minutes...

For medical applications, you need to use isotopes that have a suitable half-life. A radioactive tracer needs to have a short half-life to minimise the risk of damage to the patient. A source for sterilising equipment needs to have a long half-life, so you don't have to replace it too often (see next page).

Uses of Radiation

Radiation gets a lot of bad press, but the fact is it's essential for things like modern medicine. Read on chaps...

Smoke Detectors — Use α-Radiation

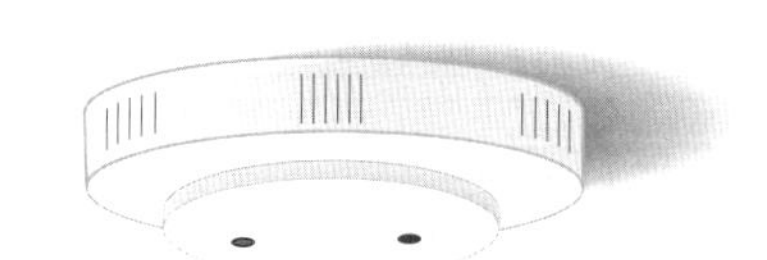

1) A weak source of alpha radiation is placed in the detector, close to two electrodes.
2) The source causes ionisation, and a current flows between the electrodes.
3) If there is a fire then smoke will absorb the radiation — so the current stops and the alarm sounds.

Tracers in Medicine — Always Short Half-Life β or γ -Emitters

1) Certain radioactive isotopes can be injected into people (or they can just swallow them) and their progress around the body can be followed using an external detector. A computer converts the reading to a display showing where the strongest reading is coming from.
2) A well-known example is the use of iodine-131, which is absorbed by the thyroid gland just like normal iodine-127, but it gives out radiation which can be detected to indicate whether the thyroid gland is taking in iodine as it should.
3) All isotopes which are taken into the body must be GAMMA or BETA emitters (never alpha), so that the radiation passes out of the body — and they should only last a few hours, so that the radioactivity inside the patient quickly disappears (i.e. they should have a short half-life).

Gamma Rays

G-M tubes Ltd.

Iodine-131 collecting in the thyroid gland

Radiotherapy — the Treatment of Cancer Using γ-Rays

1) Since high doses of gamma rays will kill all living cells, they can be used to treat cancers.
2) The gamma rays have to be directed carefully and at just the right dosage so as to kill the cancer cells without damaging too many normal cells.
3) However, a fair bit of damage is inevitably done to normal cells, which makes the patient feel very ill. But if the cancer is successfully killed off in the end, then it's worth it.

Sterilisation of Food and Surgical Instruments Using γ -Rays

1) Food can be exposed to a high dose of gamma rays which will kill all microbes, keeping the food fresh for longer.
2) Medical instruments can be sterilised in just the same way, rather than by boiling them.
3) The great advantage of irradiation over boiling is that it doesn't involve high temperatures, so things like fresh apples or plastic instruments can be totally sterilised without damaging them.
4) The food is not radioactive afterwards, so it's perfectly safe to eat.
5) The isotope used for this needs to be a very strong emitter of gamma rays with a reasonably long half-life (at least several months) so that it doesn't need replacing too often.

Ionising radiation — just what the doctor ordered...

Radiation has many important uses — especially in medicine. Make sure you know why each application uses a particular isotope according to its half-life and the type of radiation it gives out.

Radioactivity Safety

When Marie Curie discovered the radioactive properties of radium in 1898, nobody knew about its dangers. Radium was used to make glow-in-the-dark watches and many watch dial painters developed cancer as a result.

Radiation Harms Living Cells

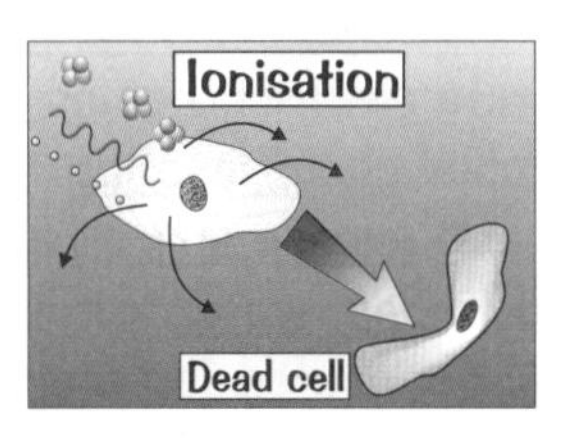

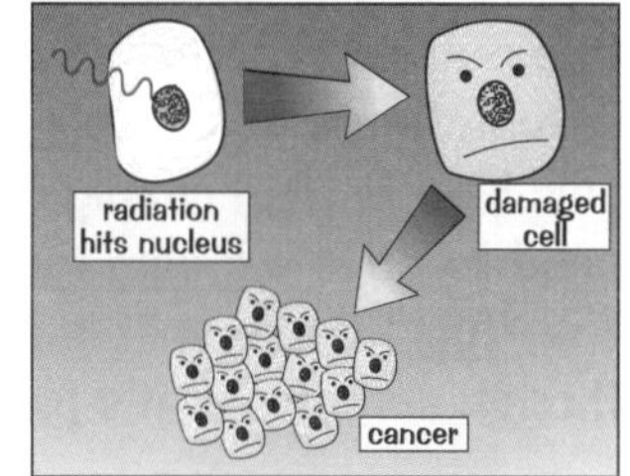

1) Alpha, beta and gamma radiation will cheerfully enter living cells and collide with molecules.
2) These collisions cause ionisation, which damages or destroys the molecules.
3) Lower doses tend to cause minor damage without killing the cell.
4) This can give rise to mutant cells which divide uncontrollably. This is cancer.
5) Higher doses tend to kill cells completely, which causes radiation sickness if a lot of body cells all get blatted at once.
6) The extent of the harmful effects depends on two things:
   a) How much exposure you have to the radiation.
   b) The energy and penetration of the radiation, since some types are more hazardous than others, of course.

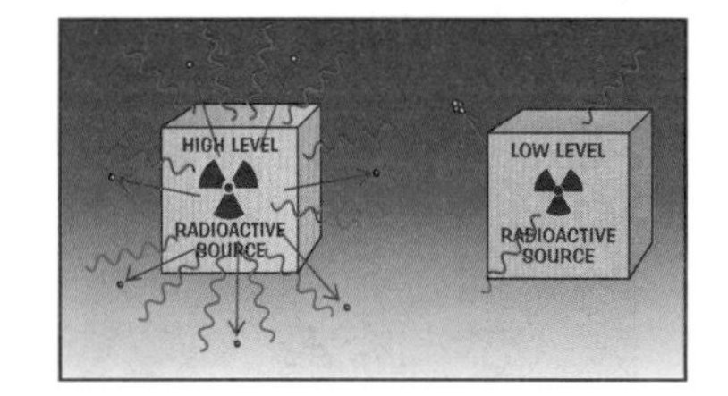

Outside the Body, β and γ–Sources are the Most Dangerous

This is because beta and gamma can get inside to the delicate organs, whereas alpha is much less dangerous because it can't penetrate the skin.

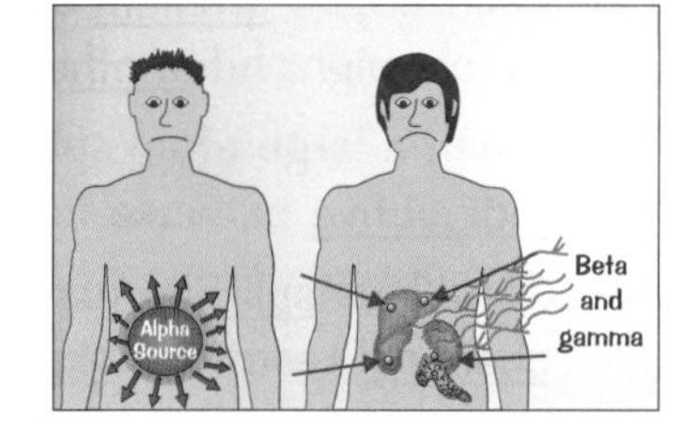

Inside the Body, an α-Source is the Most Dangerous

Inside the body alpha sources do all their damage in a very localised area. Beta and gamma sources on the other hand are less dangerous inside the body because they mostly pass straight out without doing much damage.

You Need to Learn About These Safety Precautions

Obviously radioactive materials need to be handled carefully.
But in the exam they might ask you to evaluate some specific precautions that should be taken when handling radioactive materials.

1) When conducting experiments, use radioactive sources for as short a time as possible so your exposure is kept to a minimum.
2) Never allow skin contact with a source. Always handle with tongs.
3) Hold the source at arm's length to keep it as far from the body as possible. This will decrease the amount of radiation that hits you, especially for alpha particles as they don't travel far in air.
4) Keep the source pointing away from the body and avoid looking directly at it.
5) Lead absorbs all three types of radiation (though a lot of it is needed to stop gamma radiation completely). Always store radioactive sources in a lead box and put them away as soon as the experiment is over. Medical professionals who work with radiation every day (such as radiographers) wear lead aprons and stand behind lead screens for extra protection because of its radiation absorbing properties.
6) When someone needs an X-ray or radiotherapy, only the area of the body that needs to be treated is exposed to radiation. The rest of the body is protected with lead or other radiation absorbing materials.

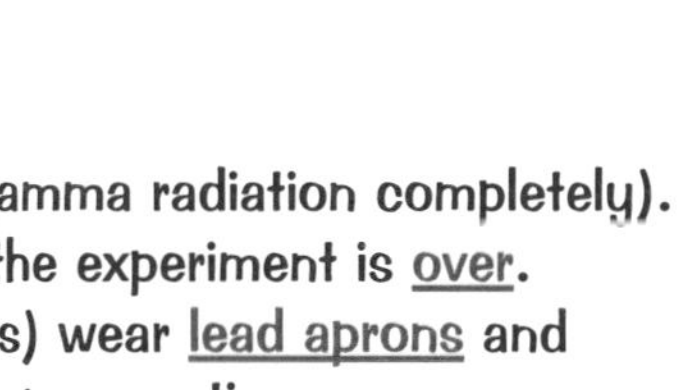

Radiation sickness — well yes, it does all get a bit tedious...

Sadly, much of our knowledge of the harmful effects of radiation has come as a result of devastating events such as the atomic bombing of Japan in 1945. In the months following the bombs, thousands suffered from radiation sickness — the symptoms of which include nausea, fatigue, skin burns, hair loss and, in serious cases, death. In the long term, the area has experienced increased rates of cancer, particularly leukaemia.

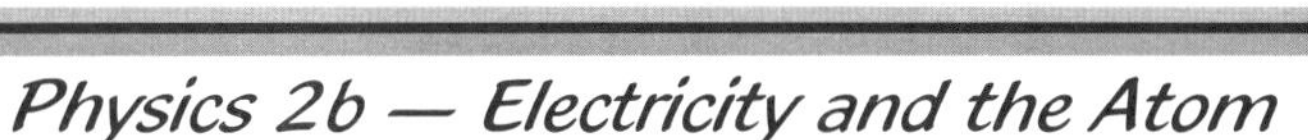

Nuclear Fission and Fusion

Unstable isotopes aren't just good for medicine — with the right set-up you can generate some serious energy.

Nuclear Fission — the Splitting Up of Big Atomic Nuclei

Nuclear power stations generate electricity using nuclear reactors. In a nuclear reactor, a controlled chain reaction takes place in which atomic nuclei split up and release energy in the form of heat. This heat is then simply used to heat water to make steam, which is used to drive a steam turbine connected to an electricity generator. The "fuel" that's split is usually uranium-235, though sometimes it's plutonium-239 (or both).

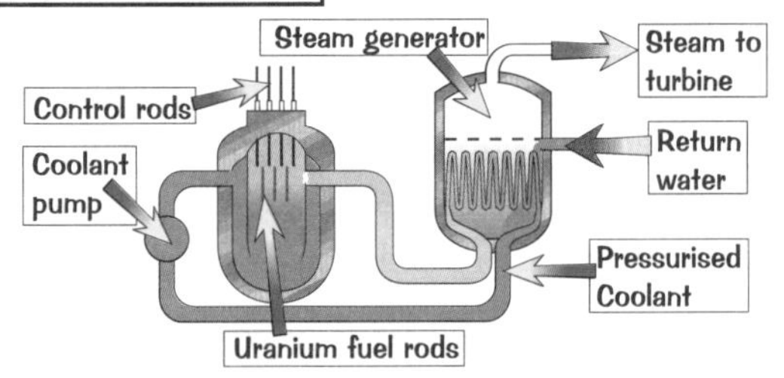

The Chain Reactions:

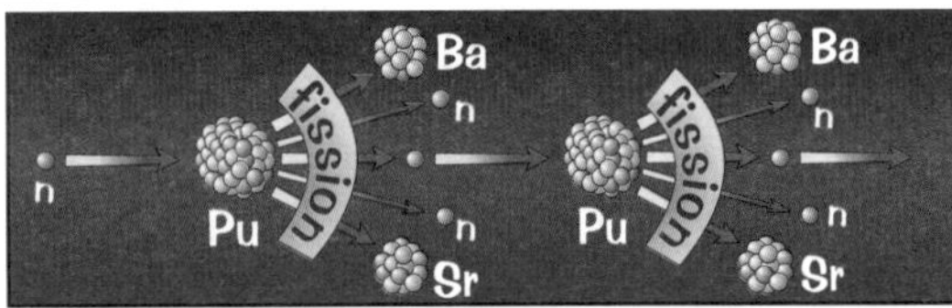

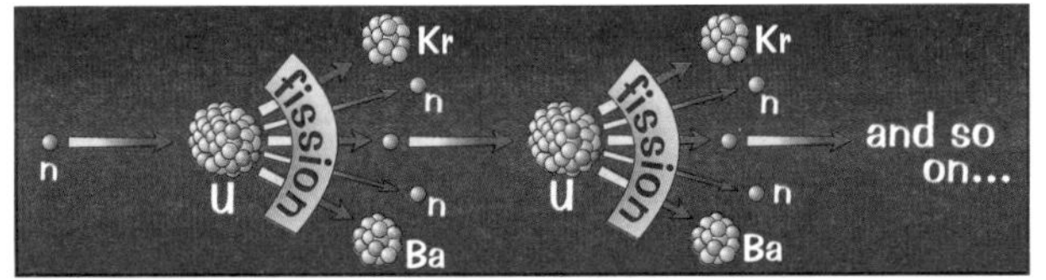

1) For nuclear fission to happen, a slow moving neutron must be absorbed into a uranium or plutonium nucleus. This addition of a neutron makes the nucleus unstable, causing it to split.
2) Each time a uranium or plutonium nucleus splits up, it spits out two or three neutrons, one of which might hit another nucleus, causing it to split also, and thus keeping the chain reaction going.
3) When a large atom splits in two it will form two new smaller nuclei. These new nuclei are usually radioactive because they have the "wrong" number of neutrons in them.
4) A nucleus splitting (called a fission) gives out a lot of energy — lots more energy than you get from any chemical reaction. Nuclear processes release much more energy than chemical processes do. That's why nuclear bombs are so much more powerful than ordinary bombs (which rely on chemical reactions).
5) The main problem with nuclear power is with the disposal of waste. The products left over after nuclear fission are highly radioactive, so they can't just be thrown away. They're very difficult and expensive to dispose of safely.
6) Nuclear fuel is cheap but the overall cost of nuclear power is high due to the cost of the power plant and final decommissioning. Dismantling a nuclear plant safely takes decades.
7) Nuclear power also carries the risk of radiation leaks from the plant or a major catastrophe like Chernobyl.

Nuclear Fusion — the Joining of Small Atomic Nuclei

1) Two light nuclei (e.g. hydrogen) can join to create a larger nucleus — this is called nuclear fusion.
2) Fusion releases a lot of energy (more than fission for a given mass) — all the energy released in stars comes from fusion (see next page). So people are trying to develop fusion reactors to generate electricity.
3) Fusion doesn't leave behind a lot of radioactive waste like fission, and there's plenty of hydrogen knocking about to use as fuel.

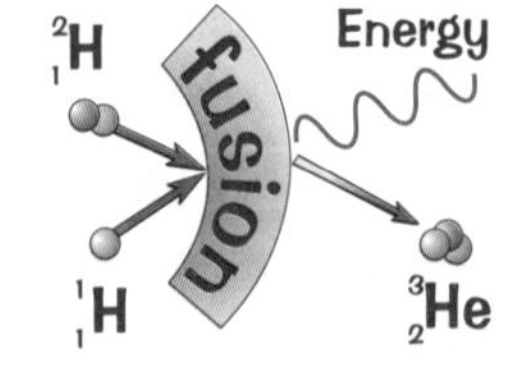

4) The big problem is that fusion can only happen at really high temperatures — about 10 000 000 °C.
5) You can't hold the hydrogen at the high temperatures and pressures required for fusion in an ordinary container — you need an extremely strong magnetic field.
6) There are a few experimental reactors around, but none of them are generating electricity yet. At the moment it takes more power to get up to temperature than the reactor can produce.

Ten million degrees — that's hot...

It'd be great if we could get nuclear fusion to work — there's loads of fuel available and it doesn't create much radioactive waste compared with fission. It's a shame that at the moment we need to use more energy to create the conditions for fusion than we can get out of it. Make sure you know the pros and cons of fission and fusion.

The Life Cycle of Stars

Stars go through many traumatic stages in their lives — just like teenagers.

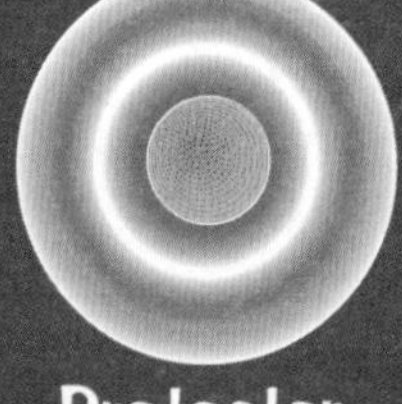

Protostar

1) Stars initially form from clouds of DUST AND GAS. The force of gravity makes the gas and dust spiral in together to form a protostar.

2) Gravitational energy is converted into heat energy, so the temperature rises. When the temperature gets high enough, hydrogen nuclei undergo nuclear fusion to form helium nuclei and give out massive amounts of heat and light. A star is born. Smaller masses of gas and dust may also pull together to make planets that orbit the star.

Main Sequence Star

3) The star immediately enters a long stable period, where the heat created by the nuclear fusion provides an outward pressure to balance the force of gravity pulling everything inwards. The star maintains its energy output for millions of years due to the massive amounts of hydrogen it consumes. In this stable period it's called a MAIN SEQUENCE STAR and it lasts several billion years. (The Sun is in the middle of this stable period — or to put it another way, the Earth has already had half its innings before the Sun engulfs it!)

Stars much bigger than the Sun

Stars about the same size as the Sun

4) Eventually the hydrogen begins to run out. Heavier elements such as iron are made by nuclear fusion of helium. The star then swells into a RED GIANT, if it's a small star, or a RED SUPER GIANT if it's a big star. It becomes red because the surface cools.

Red Giant

White Dwarf

Red Super Giant

5) A small-to-medium-sized star like the Sun then becomes unstable and ejects its outer layer of dust and gas as a PLANETARY NEBULA.

6) This leaves behind a hot, dense solid core — a WHITE DWARF, which just cools down to a BLACK DWARF and eventually disappears.

Supernova

Neutron Star...

...or Black Hole

7) Big stars, however, start to glow brightly again as they undergo more fusion and expand and contract several times, forming elements as heavy as iron in various nuclear reactions. Eventually they explode in a SUPERNOVA, forming elements heavier than iron and ejecting them into the universe to form new planets and stars.

8) The exploding supernova throws the outer layers of dust and gas into space, leaving a very dense core called a NEUTRON STAR. If the star is big enough this will become a BLACK HOLE.

Red Giants, White Dwarfs, Black Holes, Green Ghosts...

The early universe contained only hydrogen, the simplest and lightest element. It's only thanks to nuclear fusion inside stars that we have any of the other naturally occurring elements. Remember — the heaviest element produced in stable stars is iron, but it takes a supernova (or a lab) to create the rest.

Revision Summary for Physics 2b

There's some pretty heavy physics in this section. But just take it one page at a time and it's not so bad. You're even allowed to go back through the pages for a sneaky peak if you get stuck on these questions.

1) What causes the build-up of static electricity? Which particles move when static builds up?
2) True or false: the greater the resistance of an electrical component, the smaller the current that flows through it?
3)* 240 C of charge is carried though a wire in a circuit in one minute. How much current has flowed through the wire?
4) What formula relates work done, potential difference and charge?
5) Draw a diagram of the circuit that you would use to find the resistance of a motor.
6) Sketch typical potential difference-current graphs for:
   a) a resistor, b) a filament lamp, c) a diode. Explain the shape of each graph.
7) Explain how resistance of a component changes with its temperature in terms of ions and electrons.
8)* What potential difference is required to push 2 A of current through a 0.6 Ω resistor?
9) Give three applications of LEDs.
10) Describe how the resistance of an LDR varies with light intensity. Give an application of an LDR.
11)* A 4 Ω bulb and a 6 Ω bulb are connected in series with a 12 V battery.
    a) How much current flows through the 4 Ω bulb?
    b) What is the potential difference over the 6 Ω bulb?
    c) What would the potential difference over the 6 Ω bulb be if the two bulbs were connected in parallel?
12)* An AC supply of electricity has a time period of 0.08s. What is its frequency?
13) Name the three wires in a three-core cable.
14) Sketch and label a properly wired three-pin plug.
15) Explain fully how a fuse and earth wire work together.
16) How does an RCCB stop you from getting electrocuted?
17)* Which uses more energy, a 45 W pair of hair straighteners used for 5 minutes, or a 105 W hair dryer used for 2 minutes?
18)* Find the appropriate fuse (3 A, 5 A or 13 A) for these appliances:
    a) a toaster rated at 230 V, 1100 W   b) an electric heater rated at 230 V, 2000 W
19)* Calculate the energy transformed by a torch using a 6 V battery when 530 C of charge pass through.
20) Explain how the experiments of Rutherford and Marsden led to the nuclear model of the atom.
21) Draw a table stating the relative mass and charge of the three basic subatomic particles.
22) True or false: radioactive decay can be triggered by certain chemical reactions?
23) What type of subatomic particle is a beta particle?
24) List two places where the level of background radiation is increased and explain why.
25) Name three occupations that have an increased risk of exposure to radiation.
26) Sketch the paths of an alpha particle and a beta particle travelling through an electric field.
27) What is the definition of half-life?
28) Give an example of how gamma radiation can be used in medicine.
29) Which is the most dangerous form of radiation if you eat it? Why?
30) Describe the precautions you should take when handling radioactive sources in the laboratory.
31) Draw a diagram to illustrate the fission of uranium-235 and explain how the chain reaction works.
32) What is the main environmental problem associated with nuclear power?
33) What is nuclear fusion? Why is it difficult to construct a working fusion reactor?
34) Describe the steps that lead to the formation of a main sequence star (like our Sun).
35) Why will our Sun never form a black hole?

*Answers on page 108.

X-rays in Medicine

X-rays are ionising — they can damage living cells (see p.77) but they can be really useful if handled carefully...

X-ray Images are Used in Hospitals for Medical Diagnosis

1) X-rays are high frequency, short wavelength electromagnetic waves (see p.37). Their wavelength is roughly the same size as the diameter of an atom.
2) They are transmitted by (pass through) healthy tissue, but are absorbed by denser materials like bones and metal.
3) They affect photographic film in the same way as light, which means they can be used to take photographs.
4) X-ray photographs can be used to diagnose many medical conditions such as bone fractures or dental problems (problems with your teeth).
5) X-ray images can be formed electronically using charge-coupled devices (CCDs). CCDs are silicon chips about the size of a postage stamp, divided up into a grid of millions of identical pixels. CCDs detect X-rays and produce electronic signals which are used to form high resolution images. The same technology is used to take photographs in digital cameras.

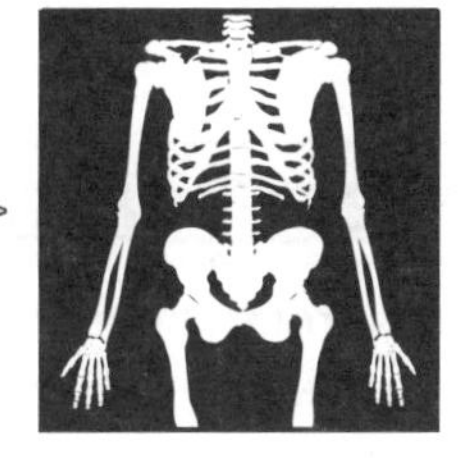

The brighter bits are where fewer X-rays get through. This is a negative image. The plate starts off all white.

CT Scans use X-rays

Computerised axial tomography (CT) scans use X-rays to produce high resolution images of soft and hard tissue. The patient is put inside the cylindrical scanner, and an X-ray beam is fired through the body from an X-ray tube and picked up by detectors on the opposite side. The X-ray tube and detectors are rotated during the scan. A computer interprets the signals from the detectors to form an image of a two-dimensional slice through the body. Multiple two-dimensional CT scans can be put together to make a three-dimensional image of the inside of the body.

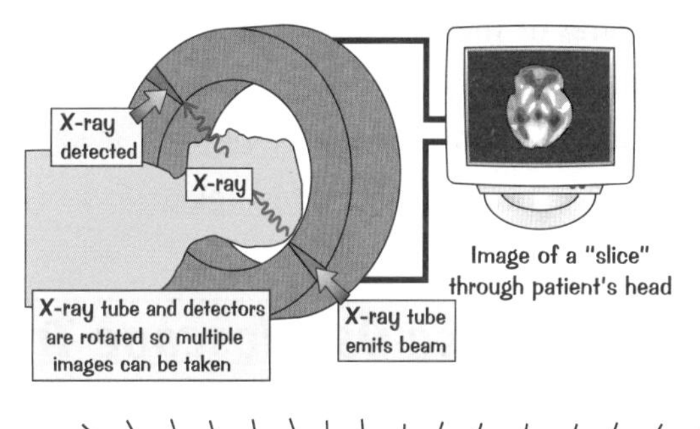

Soft tissue can absorb a small amount of X-ray radiation. CT scans use lots of X-rays (more than normal X-ray photographs) to distinguish between the tiny variations in tissue density.

X-rays can be Used to Treat Cancer

X-rays can cause ionisation — high doses of X-rays will kill living cells. They can therefore be used to treat cancers, just like gamma radiation (see p.76). The X-rays have to be carefully focused and at just the right dosage to kill the cancer cells without damaging too many normal cells.

TO TREAT CANCER:

1) The X-rays are focused on the tumour using a wide beam.
2) This beam is rotated round the patient with the tumour at the centre.
3) This minimises the exposure of normal cells to radiation, and so reduces the chances of damaging the rest of the body.

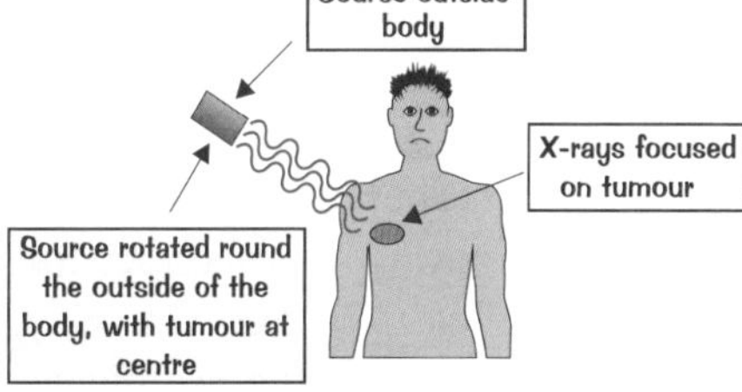

Radiographers Take Precautions to Minimise Radiation Dose

Prolonged exposure to ionising radiation can be very dangerous to your health.

1) Radiographers who work with X-ray machines or CT scanners need to take precautions to minimise their X-ray dose (see p.74).
2) They wear lead aprons, stand behind a lead screen, or leave the room while scans are being done.
3) Lead is used to shield areas of the patient's body that aren't being scanned, and the exposure time to the X-rays is always kept to an absolute minimum.

Don't just scan this page — focus on it...

As well as having lead aprons and screens to stand behind, radiographers wear special badges that record the amount of radiation they are exposed to. This means that their radiation dose can be monitored and regulated.

Ultrasound

There's sound, and then there's ultrasound.

Ultrasound is Sound with a Higher Frequency than We Can Hear

Electrical systems can be made which produce electrical oscillations of any frequency.
These can easily be converted into mechanical vibrations to produce sound waves of a higher frequency than the upper limit of human hearing (the range of human hearing is 20 to 20 000 Hz). This is called ultrasound.

Ultrasound Waves Get Partially Reflected at a Boundary Between Media

1) When a wave passes from one medium into another, some of the wave is reflected off the boundary between the two media, and some is transmitted (and refracted). This is partial reflection.
2) What this means is that you can point a pulse of ultrasound at an object, and wherever there are boundaries between one substance and another, some of the ultrasound gets reflected back.
3) The time it takes for the reflections to reach a detector can be used to measure how far away the boundary is.
4) This is how ultrasound imaging works (see p.83).

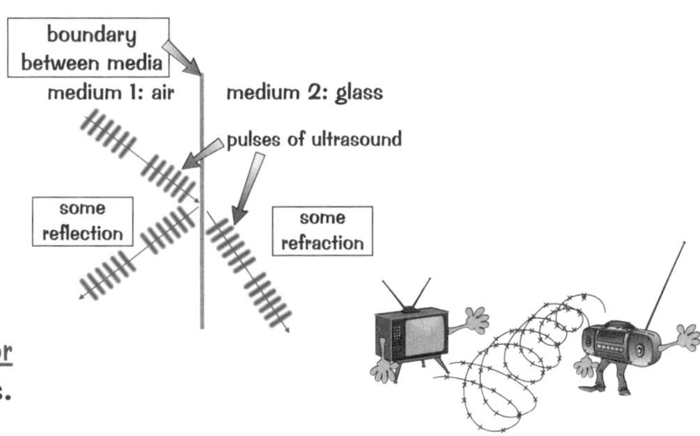

You Can Use Oscilloscope Traces to Find Boundaries

1) The oscilloscope trace on the right shows an ultrasound pulse reflecting off two separate boundaries.
2) Given the "seconds per division" setting of the oscilloscope (see p.66), you can work out the time between the pulses by measuring on the screen.
3) If you know the speed of sound in the medium, you can work out the distance between the boundaries, using this formula:
s is distance in metres, m.
v is speed in metres per second, m/s.
t is time in seconds, s.

$$s = v \times t$$

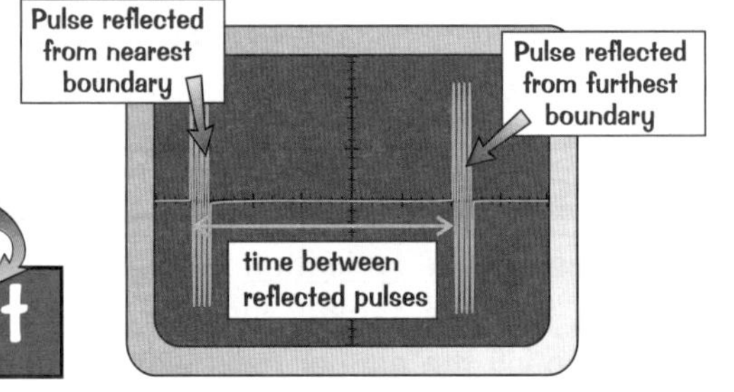

Example: A pulse of ultrasound is beamed into a patient's abdomen. The first boundary it reflects off is between fat and muscle. The second boundary is between muscle and a body cavity. An oscilloscope trace shows that the time between the reflected pulses is 10 μs. The ultrasound travels at a speed of 1500 m/s. Calculate the distance between the fat/muscle boundary and the muscle/cavity boundary.

1 μs = 0.000001 s

So, you'll need to find the distance using $s = v \times t$.
BUT, the reflected pulses have travelled there and back, so the distance you calculate will be twice the distance between boundaries (think about it).
$s = v \times t = 1500 \times 0.00001 = 0.015$ m.
So the distance between boundaries $= 0.015 \div 2 = 0.0075$ m = 7.5 mm.

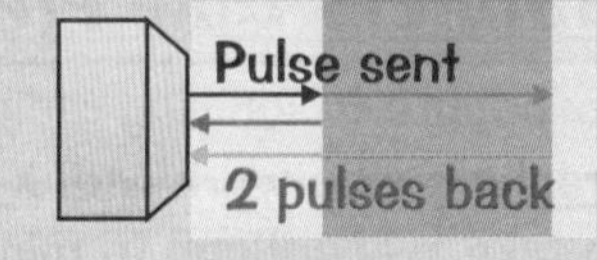

Partially reflected — completely revised...

It's crazy to think that you can use sound waves to make a image — but that's the basis of ultrasound scanning (see next page). And it all comes down to some simple reflection and a bit of distance = speed × time.

Ultrasound Uses

Those high frequency sound waves aren't just for playing with oscilloscopes, you know.
They actually happen to be well useful, especially when people are ill or pregnant.

Ultrasound Waves can be Used in Medicine

Ultrasound has a variety of uses in medicine, from investigating blood flow in organs, to diagnosing heart problems, to checking on fetal development. The examples below are two of the most common. Read on...

Breaking Down Kidney Stones

Kidney stones are hard masses that can block the urinary tract — ouch. An ultrasound beam concentrates high-energy waves at the kidney stone and turns it into sand-like particles. These particles then pass out of the body in the urine. It's a good method because the patient doesn't need surgery and it's relatively painless.

Pre-Natal Scanning of a Fetus

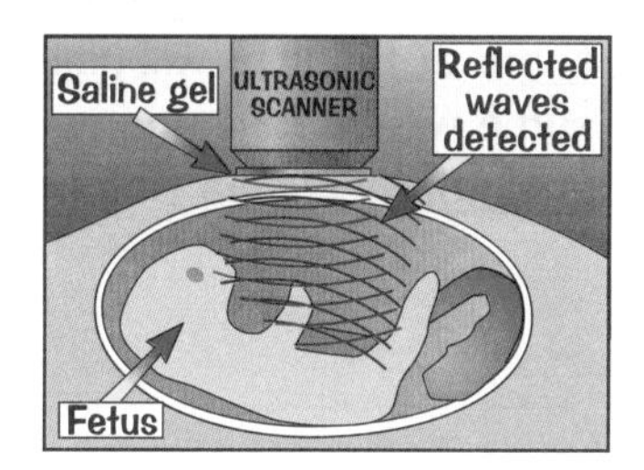

Ultrasound waves can pass through the body, but whenever they reach a boundary between two different media (like fluid in the womb and the skin of the fetus) some of the wave is reflected back and detected (see p.82). The exact timing and distribution of these echoes are processed by a computer to produce a video image of the fetus.

Medical Imaging is Full of Compromises...

Doctors have to make a compromises between getting a good enough image to be able to diagnose problems, whilst putting the patient at as low a risk as possible. X-ray and ultrasound imaging both have their advantages and disadvantages...

IS IT SAFE?

1) Ultrasound waves are non-ionising and, as far as anyone can tell, safe.
2) X-rays are ionising. They can cause cancer if you're exposed to too high a dose, and are definitely NOT safe to use on developing babies (see p.77).
3) CT scans use a lot more X-ray radiation than standard X-ray photographs, so the patient is exposed to even more ionising radiation. Generally CT scans aren't taken unless they are really needed because of the increased radiation dose.

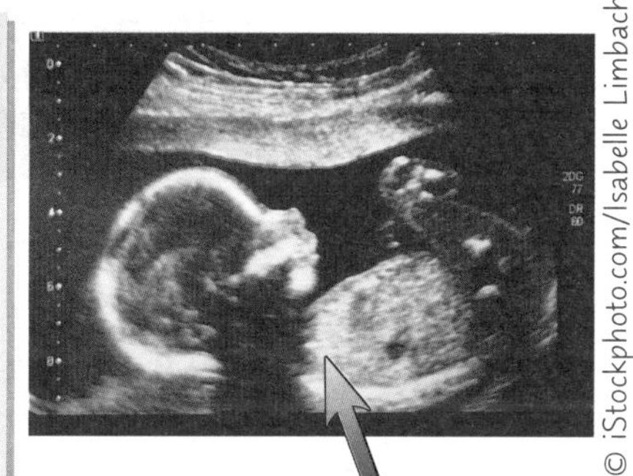

Ultrasound scans are safe for the fetus, but they do give a fuzzy image.

WHAT ABOUT IMAGE QUALITY?

1) Ultrasound images are typically fuzzy — which can make it harder to diagnose some conditions using these images.
2) X-ray photographs produce clear images of bones and metal, but not a lot else.
3) CT scans produce detailed images and can be used to diagnose complicated illnesses, as the high resolution images can make it easier to work out the problem. High quality 3D images can also be used in the planning of complicated surgery.

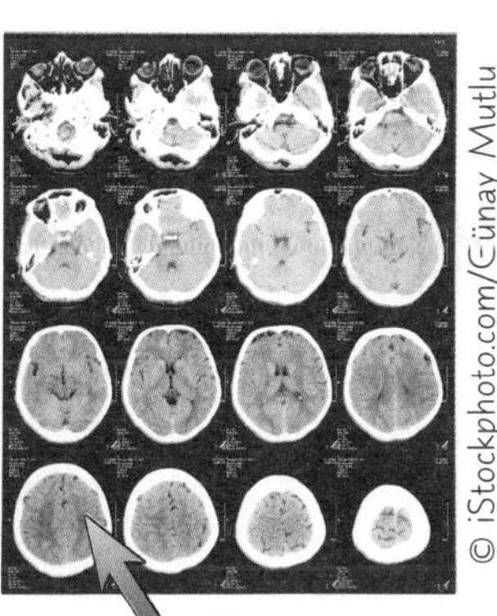

CT scans of the brain are very detailed and clear.

What did the X-ray say to the muscle? Just passing through...

It's important that you're able to compare and contrast the different imaging techniques. Sit down with a pen and paper and write yourself a mini essay all about medical imaging and all the advantages and disadvantages.

Refractive Index

You might remember refraction from way back in the day when you did Physics 1b (see p.36). Here it is again in all its glory (but with some maths thrown in for the laughs).

Refraction is Caused by the Waves Changing Speed

Refraction is when waves change direction as they enter a different medium. This is caused by the change in density from one medium to the other — which changes the speed of the waves.

1) When waves slow down they bend towards the normal.
2) When light enters glass or plastic it slows down — to about 2/3 of its speed in air.
3) If a wave hits a boundary at 90° (i.e. along the normal) it will not change direction — but it'll still slow down.
4) When light hits a different medium (e.g. plastic or glass) some of the light will pass through the new medium but some will be reflected — it all depends on the angle of incidence (the angle it hits the medium).

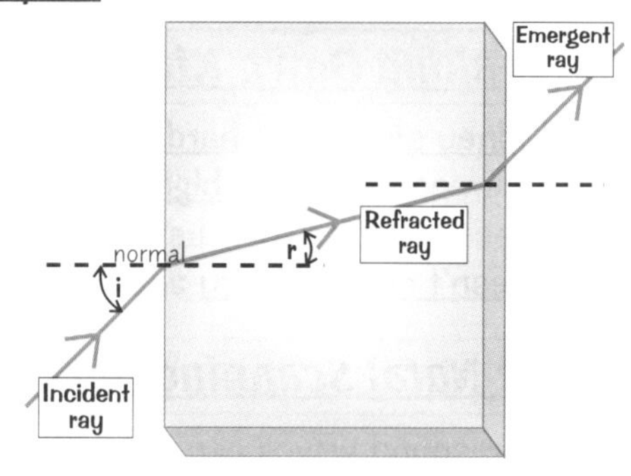

Every Transparent Material Has a Refractive Index

1) Refractive index of a medium is the ratio of speed of light in a vacuum to speed of light in that medium.
2) The angle of incidence, i, angle of refraction, r, and refractive index, n, are all linked.
3) When an incident ray passes from air into another material, the angle of refraction of the ray depends on the refractive index of the material:

$$\text{refractive index } (n) = \frac{\sin i}{\sin r}$$

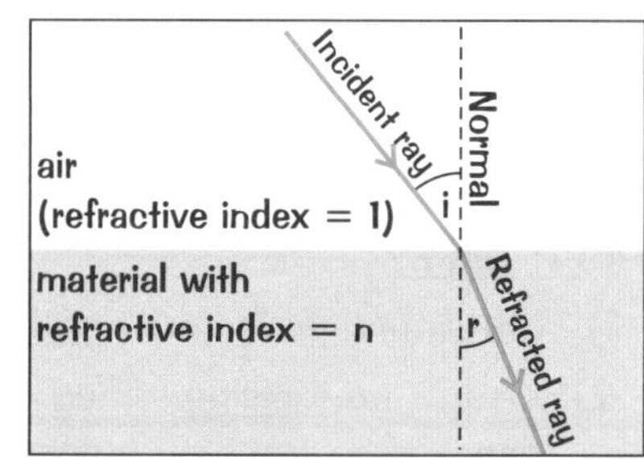

4) So if you know any two of n, i or r, you can work out the missing one.

Example 1

Jacob does an experiment to find out the refractive index of his strawberry flavour jelly. He finds that when the angle of incidence for a light beam travelling into his jelly is 42°, the angle of refraction is 35°. What is the refractive index of this particular type of jelly?

$n = \frac{\sin i}{\sin r}$ $\quad \sin i = \sin 42 = 0.67$, $\sin r = \sin 35 = 0.57$ $\Rightarrow$ so $n_{jelly} = \frac{0.67}{0.57} = 1.18$

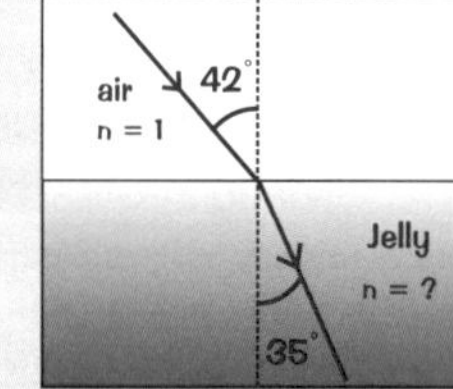

Example 2

A beam of light travels from air into water (refractive index n = 1.33). The angle of incidence is 23°. Calculate the angle of refraction to the nearest degree.

$\sin r = \frac{\sin i}{n} = \frac{\sin 23}{1.33} = 0.29 \Rightarrow$ so $r = \sin^{-1}(0.29) = 17°$

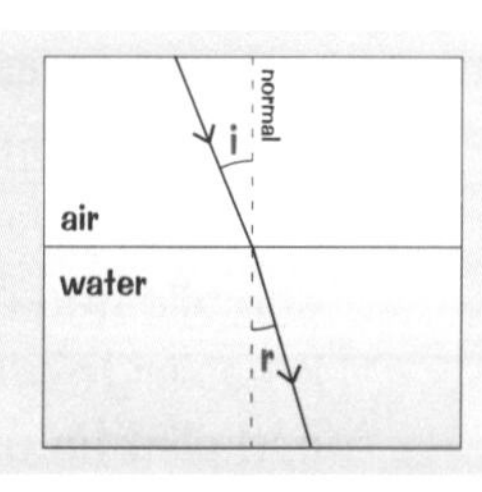

All that glitters has a high refractive index...

Make sure you get the formula for refractive index memorised and know how to use it properly. Also, it's really important that you remember that the angles of incidence and refraction are always measured from the normal to the light ray, not from the boundary between materials to the light ray. Got it? Good.

Lenses and Images

This bit is about how light acts when it hits a lens. Be ready for lots of diagrams on the next few pages.

Different Lenses Produce Different Kinds of Image

Lenses form images by refracting light and changing its direction. There are two main types of lens — converging and diverging. They have different shapes and have opposite effects on light rays.

1) A converging lens is convex — it bulges outwards. It causes parallel rays of light to converge (move together) at the principal focus.
2) A diverging lens is concave — it caves inwards. It causes parallel rays of light to diverge (spread out).
3) The axis of a lens is a line passing through the middle of the lens.
4) The principal focus of a converging lens is where rays hitting the lens parallel to the axis all meet.
5) The principal focus of a diverging lens is the point where rays hitting the lens parallel to the axis appear to all come from — you can trace them back until they all appear to meet up at a point behind the lens.
6) There is a principal focus on each side of the lens. The distance from the centre of the lens to the principal focus is called the focal length.

There are Three Rules for Refraction in a Converging Lens...

1) An incident ray parallel to the axis refracts through the lens and passes through the principal focus on the other side.
2) An incident ray passing through the principal focus refracts through the lens and travels parallel to the axis.
3) An incident ray passing through the centre of the lens carries on in the same direction.

See next page for more on this.

... And Three Rules for Refraction in a Diverging Lens

1) An incident ray parallel to the axis refracts through the lens, and travels in line with the principal focus (so it appears to have come from the principal focus).
2) An incident ray passing through the lens towards the principal focus refracts through the lens and travels parallel to the axis.
3) An incident ray passing through the centre of the lens carries on in the same direction.

See next page for more on this.

The neat thing about these rules is that they allow you to draw ray diagrams without bending the rays as they go into the lens and as they leave the lens. You can draw the diagrams as if each ray only changes direction once, in the middle of the lens (see next page).

Lenses can Produce Real and Virtual Images

1) A real image is where the light from an object comes together to form an image on a 'screen' — like the image formed on an eye's retina (the 'screen' at the back of an eye).
2) A virtual image is when the rays are diverging, so the light from the object appears to be coming from a completely different place.

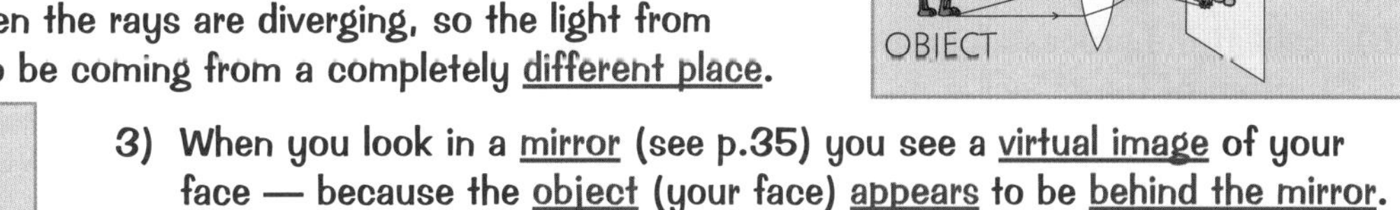

3) When you look in a mirror (see p.35) you see a virtual image of your face — because the object (your face) appears to be behind the mirror.
4) You can get a virtual image when looking at an object through a magnifying lens (see p.87) — the virtual image looks bigger than the object actually is.

To describe an image properly, you need to say 3 things: 1) How big it is compared to the object; 2) Whether it's upright or inverted (upside down) relative to the object; 3) Whether it's real or virtual.

You're virtually finished — but not really...

Remember that the proper word to describe an upside down image is inverted — you will be expected to know it.

Lenses

You might have to draw a ray diagram of refraction through a lens. Follow the instructions very carefully...

Draw a Ray Diagram for an Image Through a Converging Lens

1) Pick a point on the top of the object. Draw a ray going from the object to the lens parallel to the axis of the lens.
2) Draw another ray from the top of the object going right through the middle of the lens.
3) The incident ray that's parallel to the axis is refracted through the principal focus (F). Draw a refracted ray passing through the principal focus.
4) The ray passing through the middle of the lens doesn't bend.
5) Mark where the rays meet. That's the top of the image.
6) Repeat the process for a point on the bottom of the object. When the bottom of the object is on the axis, the bottom of the image is also on the axis.

In ray diagrams, this represents a convex lens...

object

F

object

F

image

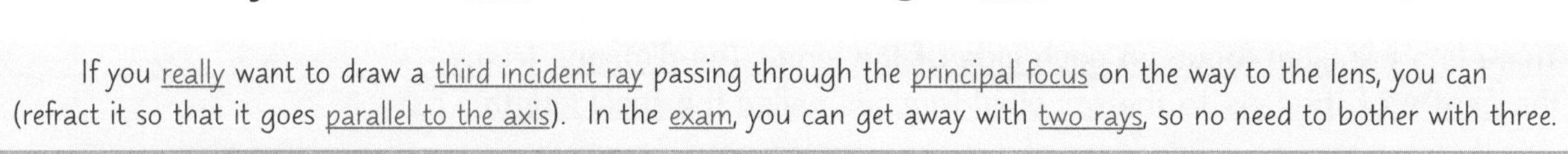
If you really want to draw a third incident ray passing through the principal focus on the way to the lens, you can (refract it so that it goes parallel to the axis). In the exam, you can get away with two rays, so no need to bother with three.

Distance from the Lens Affects the Image

1) An object at 2F will produce a real, inverted (upside down) image the same size as the object, and at 2F.

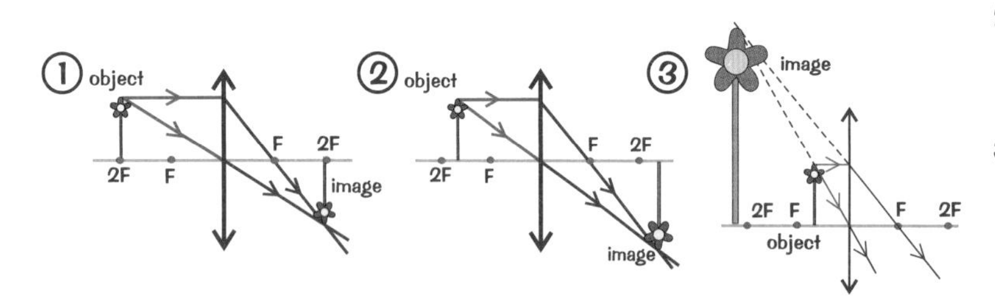

2) Between F and 2F it'll make a real, inverted image bigger than the object, and beyond 2F.
3) An object nearer than F will make a virtual image the right way up, bigger than the object, on the same side of the lens.

Draw a Ray Diagram for an Image Through a Diverging Lens

...and this represents a concave lens.

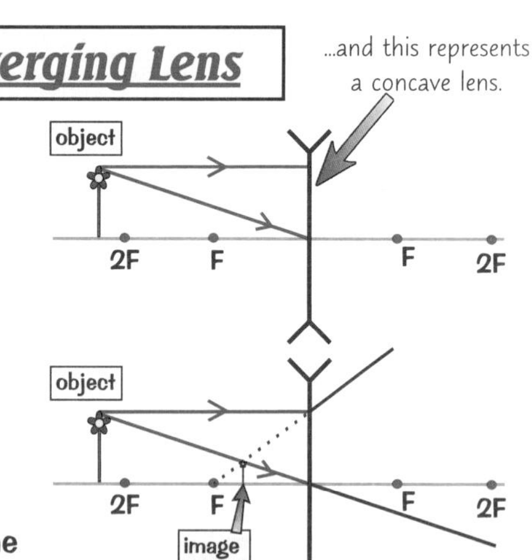

1) Pick a point on the top of the object. Draw a ray going from the object to the lens parallel to the axis of the lens.
2) Draw another ray from the top of the object going right through the middle of the lens.
3) The incident ray that's parallel to the axis is refracted so it appears to have come from the principal focus. Draw a ray from the principal focus. Make it dotted before it reaches the lens.
4) The ray passing through the middle of the lens doesn't bend.
5) Mark where the refracted rays meet. That's the top of the image.
6) Repeat the process for a point on the bottom of the object. When the bottom of the object is on the axis, the bottom of the image is also on the axis.

Again, if you really want to draw a third incident ray in the direction of the principal focus on the far side of the lens, you can. Remember to refract it so that it goes parallel to the axis. In the exam, you can get away with two rays. Choose whichever two are easiest to draw — don't try to draw a ray that won't actually pass through the lens.

The Image is Always Virtual

A diverging lens always produces a virtual image. The image is right way up, smaller than the object and on the same side of the lens as the object — no matter where the object is.

Warning — too much revision can cause attention to diverge...

Get busy practising drawing those ray diagrams. Like riding a bike, you learn by doing — so jump on it.

Magnification and Power

Converging lenses are used in magnifying glasses and in cameras. Useful for crime-fighting detectives then.

Magnifying Glasses Use Converging Lenses

Magnifying glasses work by creating a magnified virtual image (see p.85).

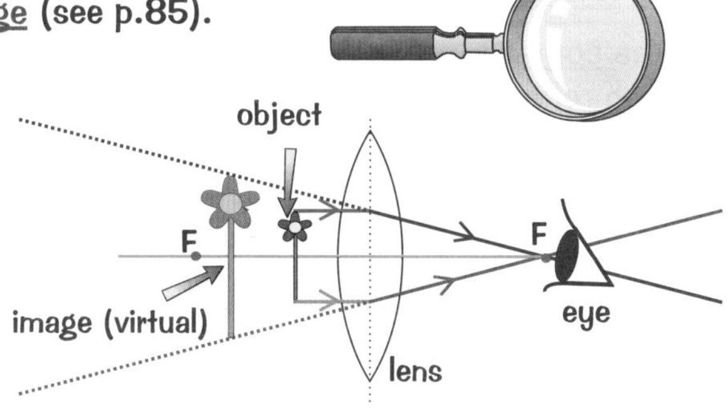

1) The object being magnified must be closer to the lens than the focal length.
2) Since the image produced is a virtual image, the light rays don't actually come from the place where the image appears to be.
3) Remember "you can't project a virtual image onto a screen" — that's a useful phrase to use in the exam if they ask you about virtual images.

Learn the Magnification Formula

You can use the magnification formula to work out the magnification produced by a lens at a given distance:

$$\text{Magnification} = \frac{\text{image height}}{\text{object height}}$$

Example: A coin with diameter 14 mm is placed a certain distance behind a magnifying lens. The virtual image produced has a diameter of 35 mm. What is the magnification of the lens at this distance?

magnification = 35 ÷ 14
= 2.5

In the exam you might have to draw a ray diagram to show where an image would be, and then measure the image so that you can work out the magnification of the lens or mirror. Another reason to draw those ray diagrams carefully...

A Powerful Lens has a Short Focal Length

1) Focal length is related to the power of the lens. The more powerful the lens, the more strongly it converges rays of light, so the shorter the focal length (see p.85).
2) The power of a lens is given by the formula:

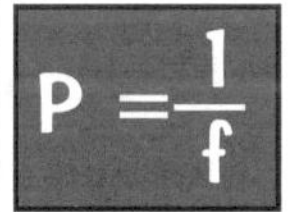

E.g. for a lens with focal length f = 0.2 m,
power = 1 ÷ 0.2 = 5 D.
(D stands for dioptres, the unit for lens power.)

3) For a converging lens, the power is positive. For a diverging lens, the power is negative.
4) The focal length of a lens is determined by two factors:

   a) the refractive index of the lens material,
   b) the curvature of the two surfaces of the lens.

5) To make a more powerful lens from a certain material like glass, you just have to make it with more strongly curved surfaces.
6) For a given focal length, the greater the refractive index of the material used to make the lens, the flatter the lens will be.
7) This means powerful lenses can be made thinner by using materials with high refractive indexes (see p.84).

He's magnificent, that pug...

People with bad eyesight used to have really thick glasses back in the day. Thankfully, nowadays you can get thin high-index plastic lenses that not only look better, but are also much more comfortable to wear. Huzzah.

The Eye

The eye is an absolute marvel of evolution — the way all the different parts work together to form an image and transport it to your brain is quite astonishing... Well, I like it anyway.

You Need to Know the Basic Structure of the Eye

1) The cornea is a transparent 'window' with a convex shape, and a high refractive index. The cornea does most of the eye's focusing.
2) The iris is the coloured part of the eye. It's made up of muscles that control the size of the pupil — the hole in the middle of the iris. This controls the intensity of light entering the eye.
3) The lens changes shape to focus light from objects at varying distances. It's connected to the ciliary muscles by the suspensory ligaments and when the ciliary muscles contract, tension is released and the lens takes on a fat, more spherical shape. When they relax, the suspensory ligaments pull the lens into a thinner, flatter shape.
4) Images are formed on the retina, which is covered in light-sensitive cells. These cells detect light and send signals to the brain to be interpreted.

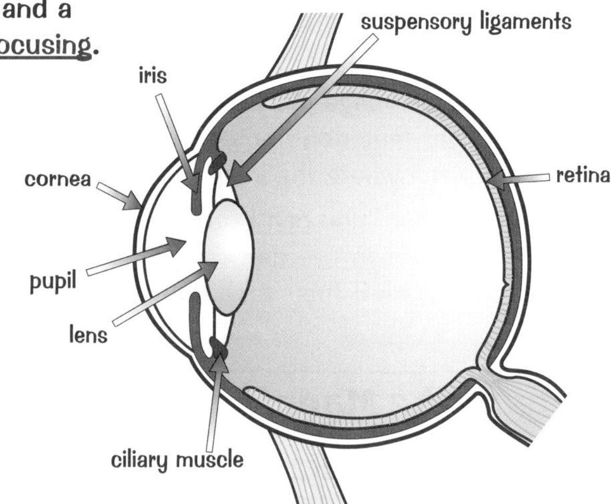

The Eye can Focus on Objects Between the Near and Far Points

1) The far point is the furthest distance that the eye can focus comfortably. For normally-sighted people, that's infinity.
2) The near point is the closest distance that the eye can focus on. For adults, the near point is approximately 25 cm.
3) As the eye focuses on closer objects, its power increases — the lens changes shape and the focal length decreases. But the distance between the lens and the image stays the same.

A Camera Forms Images in a Similar Way to the Eye

When you take a photograph of a flower, light from the object (flower) travels to the camera and is refracted by the lens, forming an image on the film.

1) The image on the film is a real image because light rays actually meet there.
2) The image is smaller than the object, because the object's a lot further away than the focal length of the lens.
3) The image is inverted (see p.86).

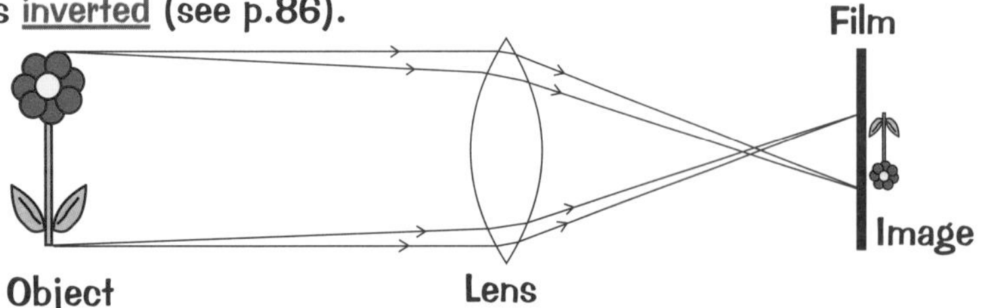

4) The same thing happens in our eye — a real, inverted image forms on the retina. Our very clever brains flip the image so that we see it right way up.
5) The film in a camera, or the CCD in a digital camera (see p.81), are the equivalent of the retina in the eye — they all detect the light focused on them and record it.

Eyes eyes baby...

The light-sensitive cells in the retina at the back of the eye send signals to the brain depending on the amount and colour of the light they've been exposed to. Then your grey matter works out all the different signals, forms an image, puts the image the right way up and figures out what it is. See, I told you it was nifty.

Correcting Vision

Sometimes things go a little awry in the eye department — and that's when physics steps in to save the day.

Short Sight is Corrected with Diverging Lenses

1) Short-sighted people can't focus on distant objects — their far point is closer than infinity (see p.88).
2) Short sight is caused by the eyeball being too long, or by the cornea and lens system being too powerful — this means the eye lens can't produce a focused image on the retina where it is supposed to.
3) Images of distant objects are brought into focus in front of the retina instead.
4) To correct short sight you need to put a diverging lens (with a negative power, see p.87) in front of the eye. This diverges light before it enters the eye, which means the lens can focus it on the retina.

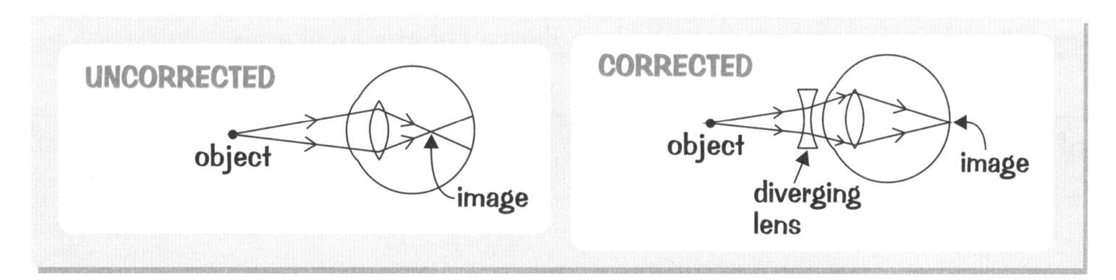

Long Sight is Corrected with Converging Lenses

Many young children are long-sighted — their lenses grow quicker than their eyeballs.

1) Long-sighted people can't focus clearly on near objects — their near point is further away than normal (25 cm or more, see p.88).
2) Long sight happens when the cornea and lens are too weak or the eyeball is too short.
3) This means that images of near objects are brought into focus behind the retina.
4) To correct long sight a converging lens (with a positive power, see p.87) can be put in front of the eye. The light is refracted and starts to converge before it enters the eye, and the image can be focused on the retina where it belongs.

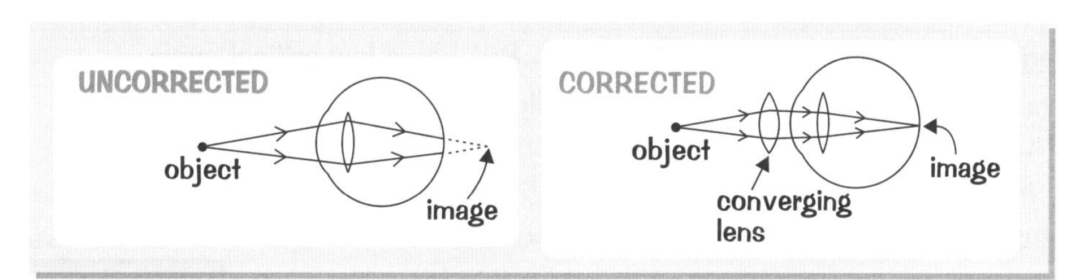

Lasers are Used to Surgically Correct Eye Problems

A laser is an narrow, intense beam of light. The light waves that come from a laser all have the same wavelength.

1) Lasers can be used in surgery to cut through body tissue, instead of using a scalpel.
2) Lasers cauterise (burn and seal shut) small blood vessels as they cut through the tissue This reduces the amount of blood the patient loses and helps to protect against infection.
3) Lasers are used to treat skin conditions such as acne scars. Lasers can be used to burn off the top layers of scarred skin revealing the less-scarred lower layers.
4) One of the most common types of laser surgery is eye surgery. A laser can be used to vaporise some of the cornea to change its shape — which changes its focusing ability. This can increase or decrease the power of the cornea so that the eye can focus images properly on the retina.

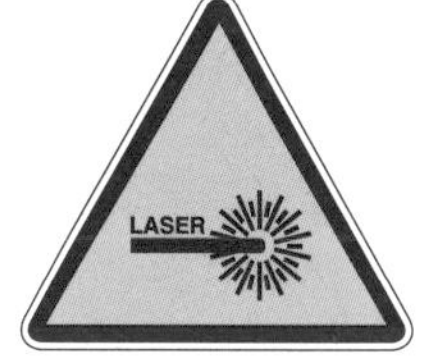

Wear glasses — they give you specs appeal...

The light from a laser can be let out in pulses for even more control during delicate eye surgery. The surgeon can precisely control how much tissue the laser takes off by using the pulses of light to do only a little bit at a time.

Total Internal Reflection

Total internal reflection is a really clever bit of physics that has loads of uses — medicine is just one of them.

Light can be Sent Along Optical Fibres Using Total Internal Reflection

1) Optical fibres can carry visible light over long distances (see p.38).
2) They work by bouncing waves off the sides of a thin inner core of glass or plastic. The wave enters one end of the fibre and is reflected repeatedly until it emerges at the other end.

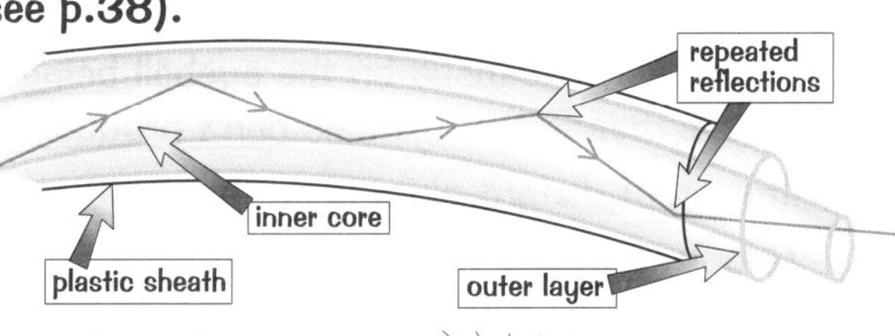

3) Optical fibres work because of total internal reflection.

4) Total internal reflection can only happen when a wave travels through a dense substance like glass or water towards a less dense substance like air.

Remember, angle of reflection, r, equals the angle of incidence, i.

5) It all depends on whether the angle of incidence is bigger than the critical angle...

If the angle of incidence (i) is...

The angle of incidence (i) and the angle of reflection (r) are always measured from the normal (a line at right angles to the surface).

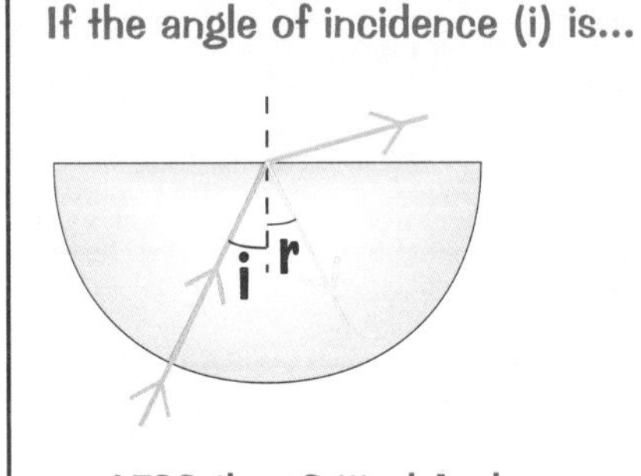

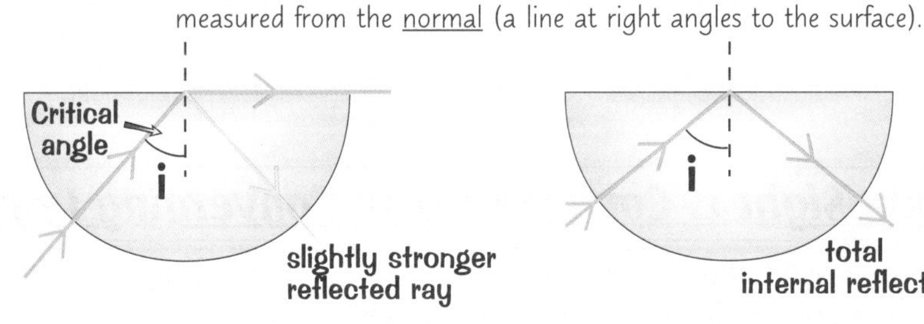

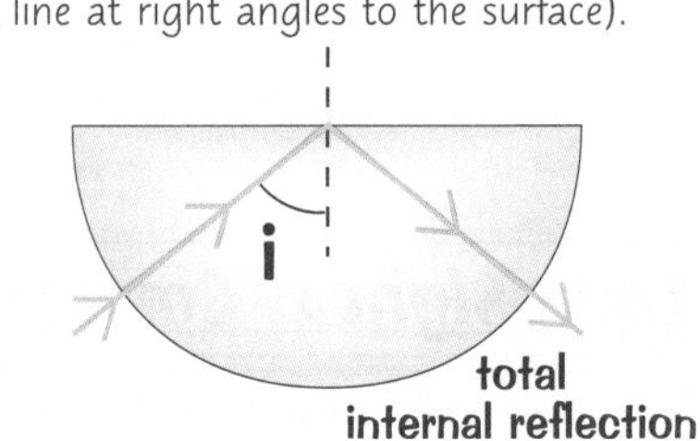

...LESS than Critical Angle:- Most of the light passes out but a little bit of it is internally reflected

...EQUAL to Critical Angle:- The emerging ray comes out along the surface. There's quite a bit of internal reflection.

...GREATER than Critical Angle:- No light comes out. It's all internally reflected, i.e. total internal reflection.

The Value of the Critical Angle Depends on the Refractive Index

1) A dense material with a high refractive index (see p.84) has a low critical angle.
2) If a material has a high refractive index, it will totally internally reflect more light — more light will be incident at an angle bigger than the critical angle.
3) For example, the critical angle of glass is around 42°, but for diamond the critical angle is just 24°, so more light is totally internally reflected — which is why diamonds are so sparkly.
4) Refractive index and critical angle (c) are related by this formula:

$$\text{Refractive index} = \frac{1}{\sin c}$$

Endoscopes Use Bundles of Optical Fibres

1) An endoscope is a thin tube containing optical fibres that lets surgeons examine inside the body.
2) Endoscopes consist of two bundles of optical fibres — one to carry light to the area of interest and one to carry an image back so that it can be viewed.

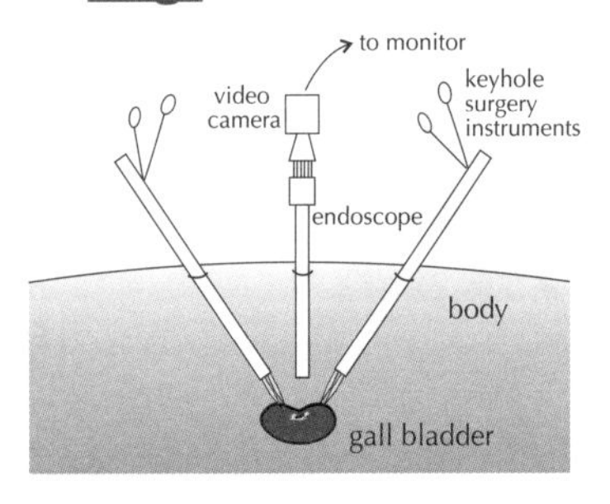

3) The image can be seen through an eyepiece or displayed as a full-colour moving image on a TV screen.
4) The big advantage of using endoscopes is that surgeons can now perform many operations by only cutting teeny holes in people — this is called keyhole surgery, and it wasn't possible before optical fibres.

Internally reflect on this a while...

Sure, endoscopes are useful for medicine, but I use mine to see when dinner's ready from the comfort of my bed...

Revision Summary for Physics 3a

Another section conquered, you absolute legend. Now all you need to do is just answer these sweet questions to see how much you've learnt. I bet it's loads.

1) What are X-rays? Name two materials X-rays are absorbed by.
2) What is a charge-coupled device?
3) How do CT scans form images of the body?
4) Describe how X-rays can be used to treat cancer.
5) What precautions can radiographers take to minimise their radiation dose?
6) What is ultrasound?
7)* Ultrasound travels through fat at a velocity of 1000 m/s. A pulse of ultrasound is sent into a person and is partially reflected off a layer of fat and a layer of muscle. The time between two reflected pulses of ultrasound is 0.00004 s. How thick is the layer of fat?
8) Explain why ultrasound rather than X-rays are used to take images of a fetus.
9) What are the advantages of using CT scans over ultrasound scans?
10) What is refraction?
11) Draw a diagram to show the path of a ray of light as it passes from air $\rightarrow$ block of glass $\rightarrow$ air, meeting the block of glass at an angle.
12)* What is the formula for refractive index? Calculate the refractive index of a block of clear plastic if a beam of light enters it with an angle of incidence of 27° and is refracted at an angle of 18°.
13) What is meant by the principal focus of a lens?
14) Draw a ray diagram for light from a distant object being focused by a convex lens. What type of image is formed?
15) What type of lenses are used to make magnifying glasses?
16)* Peter measures the length of a seed to be 1.5 cm. When he looks at the seed through a converging lens at a certain distance, the seed appears to have a length of 4.5 cm. What is the magnification of this lens at this distance?
17)* What is the power of a lens with focal length of 10 cm?
18) What two things affect the focal length of a lens?
19) Draw a labelled diagram of the eye. Describe how the eye forms a focused image on the retina.
20) What is the far point of vision?
21) What is the near point of vision? Give an approximate value of the near point for adults.
22) Describe two causes of short sight. How can short sight be corrected using lenses?
23) Describe two causes of long sight. How can long sight be corrected using lenses?
24) Describe how lasers can be used to correct vision problems.
25) a) What is total internal reflection?
    b) What happens if the angle of incidence is less than the critical angle?
    c) What happens if it is more than the critical angle?
26) a) What is an endoscope?
    b) Explain how an endoscope uses total internal reflection.
    c) Name one medical technique made possible by endoscopy.

*Answers on p.108.

Turning Forces and the Centre of Mass

Moments, they're magic. Or maybe not. Either way, expect to be royally sick of pivots by the end of the page.

A Moment is the Turning Effect of a Force

The size of the moment of the force is given by:

MOMENT = FORCE × perpendicular DISTANCE from the line of action of the force to the pivot

1) The force on the spanner causes a turning effect or moment on the nut (which acts as pivot). A larger force would mean a larger moment.

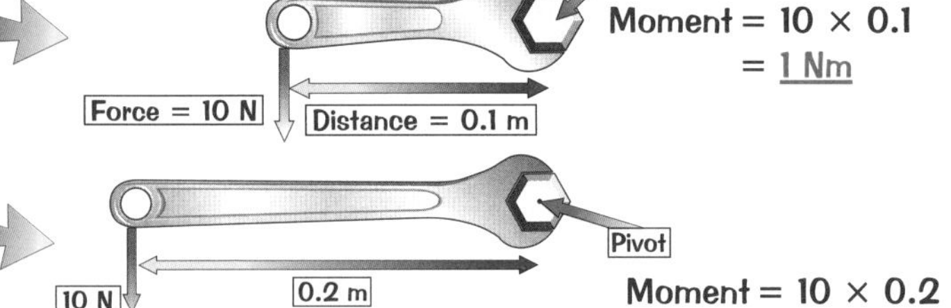

2) Using a longer spanner, the same force can exert a larger moment because the distance from the pivot is greater (see p.93).

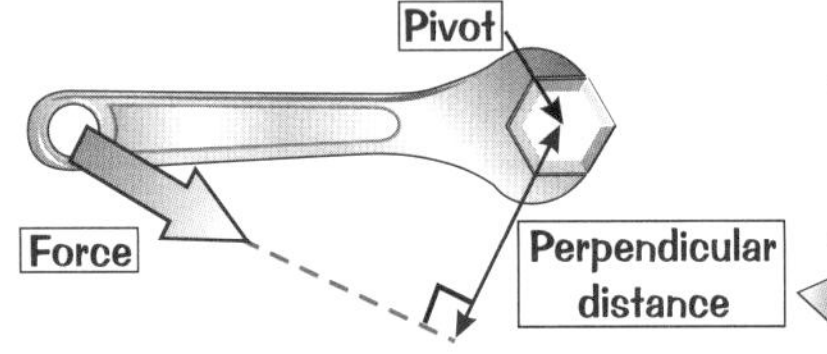

3) To get the maximum moment (or turning effect) you need to push at right angles (perpendicular) to the spanner.
4) Pushing at any other angle means a smaller moment because the perpendicular distance between the line of action and the pivot is smaller.

The Centre of Mass Hangs Directly Below the Point of Suspension

1) You can think of the centre of mass of an object as the point at which the whole mass is concentrated.
2) A freely suspended object will swing until its centre of mass is vertically below the point of suspension.

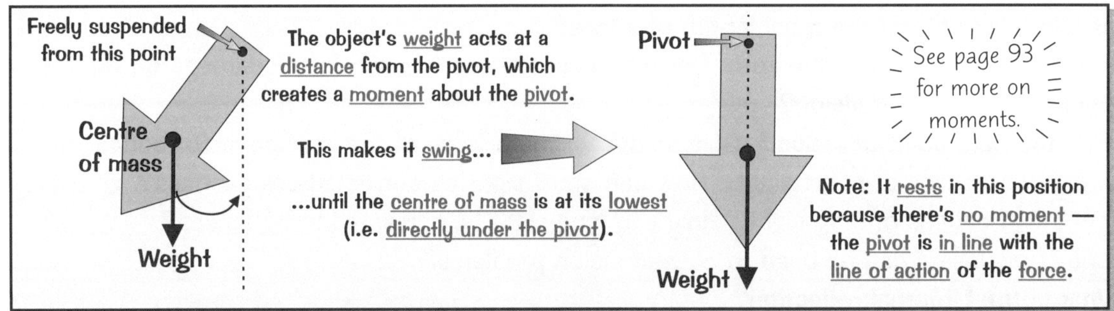

3) This means you can find the centre of mass of any flat shape like this:
   a) Suspend the shape and a plumb line from the same point, and wait until they stop moving.
   b) Draw a line along the plumb line.
   c) Do the same thing again, but suspend the shape from a different pivot point.
   d) The centre of mass is where your two lines cross.

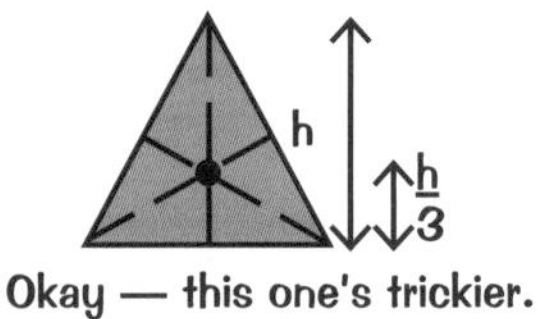

4) But you don't need to go to all that trouble for symmetrical shapes. You can quickly guess where the centre of mass is by looking for lines of symmetry.

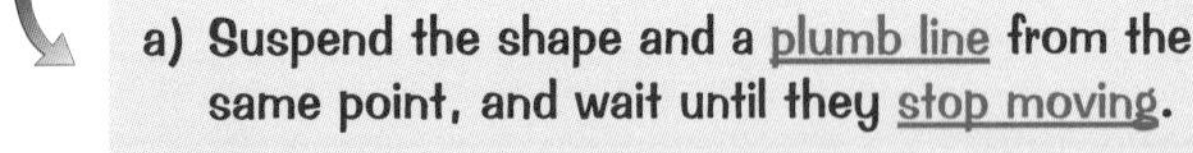

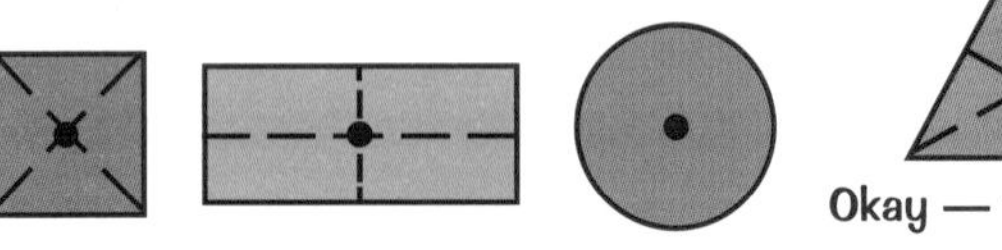

Be at the centre of mass — sit on the middle pew...

So there you go, how to find the centre of mass of your favourite piece of irregularly-shaped paper in a few easy steps. You should also now know that the next time someone asks you "How's it hanging?" your response should be "Directly below the point of suspension, thank you for asking". This page truly was an education.

Balanced Moments and Levers

Once you can calculate moments, you can work out if a seesaw is balanced. Useful thing, physics.

A Question of Balance — Are the Moments Equal?

If the anticlockwise moments are equal to the clockwise moments, the object won't turn.

Example 1: Your younger brother weighs 300 N and sits 2 m from the pivot of a seesaw. If you weigh 700 N, where should you sit to balance the seesaw?

For the seesaw to balance: **Total Anticlockwise Moments = Total Clockwise Moments**

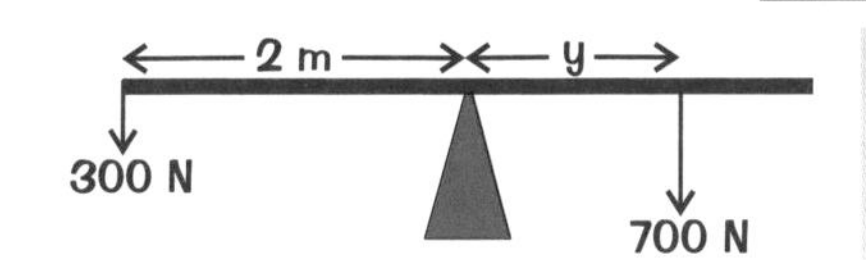

anticlockwise moment = clockwise moment

$300 \times 2 = 700 \times y$

$y = 0.86$ m

Ignore the weight of the seesaw — its centre of mass is on the pivot, so it doesn't have a turning effect.

Example 2: A 6 m long steel girder weighing 1000 N rests horizontally on a pole 1 m from one end. What is the tension in a supporting cable attached vertically to the other end?

The 'tension in the cable' bit makes it sound harder than it actually is.
But the girder's weight is balanced by the tension force in the cable, so...

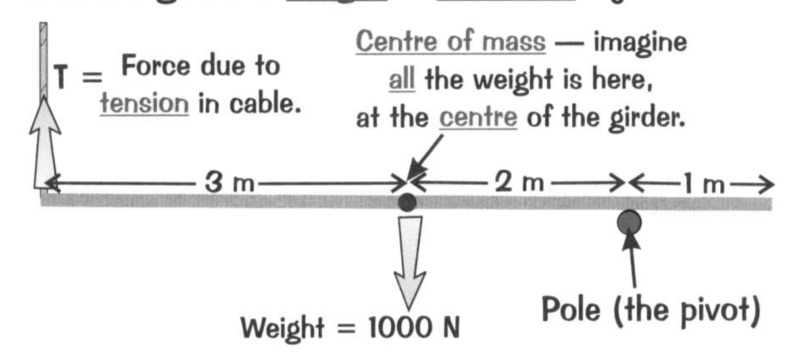

anticlockwise moment (due to weight) = clockwise moment (due to tension in cable)

$1000 \times 2 = T \times 5$

$2000 = 5T$

and so $T = 400$ N

Simple Levers use Balanced Moments

Levers use the idea of balanced moments to make it easier for us to do work (e.g. lift an object):

1) The moment needed to do work = force x distance from the pivot (see previous page). So the amount of force needed to do work depends on the distance the force is applied from the pivot.
2) Levers increase the distance from the pivot at which the force is applied — so this means less force is needed to get the same moment.
3) That's why levers are known as force multipliers — they reduce the amount of force that's needed to get the same moment by increasing the distance.

Examples of Simple Levers as Force Multipliers

Scissors use a combination of two levers.

Long sticks or bars:

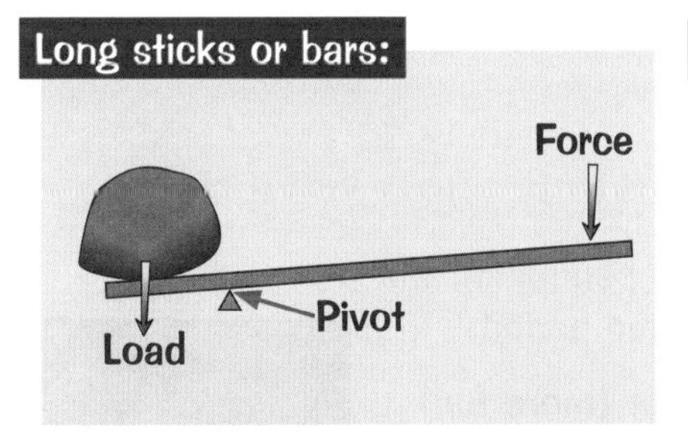

Wheelbarrows:

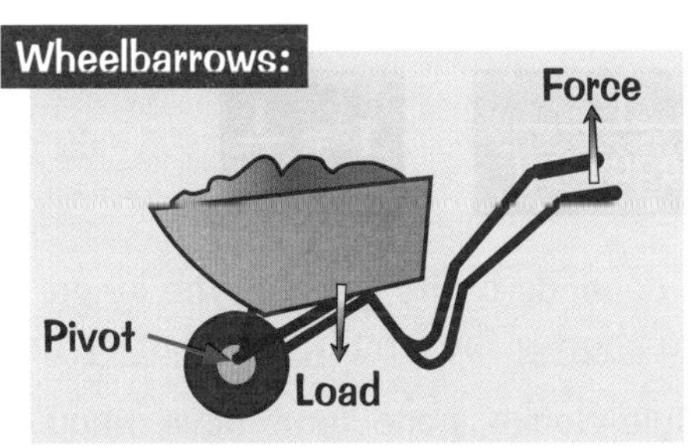

Scissors:

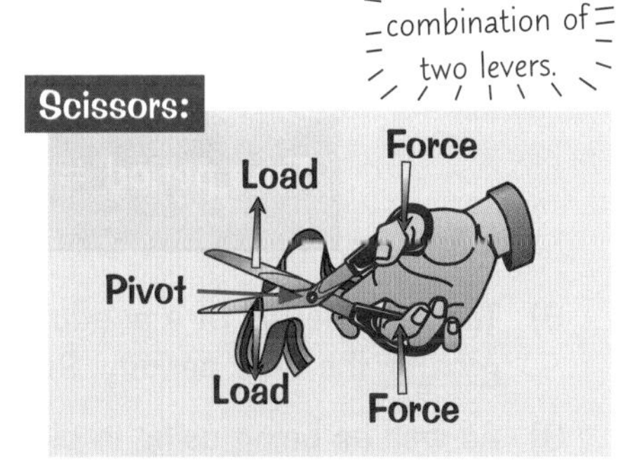

These levers make it easier to do work by moving the distance the force is applied further from the pivot.

Balanced moments — nope, not had one of those for a while...

Think of the extra force you need to open a door by pushing it near the hinge compared to at the handle — the distance from pivot is less, so you need more force to get the same moment. The best way to understand it is to do loads of practice. And learn the examples of some simple levers from above too. Good times.

Moments, Stability and Pendulums

On the last page we met total clockwise moments being balanced by total anticlockwise moments. Which is all very nice and convenient. But what happens if that isn't the case, I hear you cry. Read on my friend...

If the Moments Acting on an Object aren't Equal the Object will Turn

If the Total Anticlockwise Moments do not equal the Total Clockwise Moments, there will be a Resultant Moment ...so the object will turn.

Low and Wide Objects are Most Stable

Unstable objects tip over easily — stable ones don't. The position of the centre of mass (p.92) is all-important.

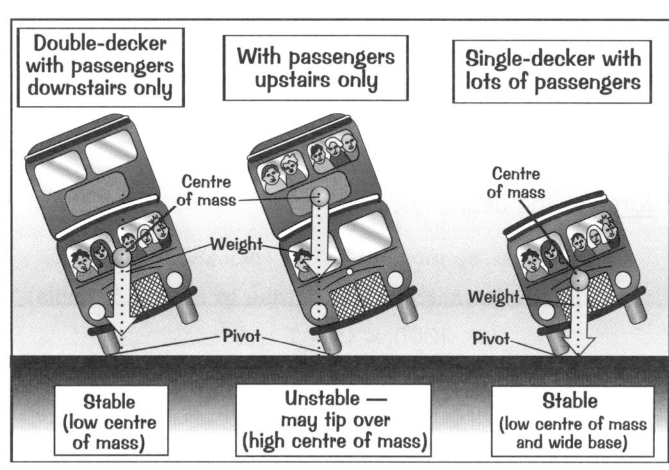

1) The most stable objects have a wide base and a low centre of mass.
2) An object will begin to tip over if its centre of mass moves beyond the edge of its base.
3) Again, it's because of moments — if the line of action of the weight of the object lies outside of the base of the object, it'll cause a resultant moment. This will tip the object over.
4) Lots of objects are specially designed to give them as much stability as possible. For example, a Bunsen burner has a wide, heavy base to give it a low centre of mass — this makes it harder to knock over.

The Time for One Pendulum Swing Depends on its Length

1) A simple pendulum is made by suspending a weight from a piece of string. When you pull back a pendulum and let it go, it will swing back and forth.
2) The time taken for the pendulum to swing from one side to the other and back again is called the time period.
3) The time period for each swing of a given pendulum is the always the same — this is what makes pendulums perfect for keeping time in clocks.
4) The time period can be calculated using this formula:

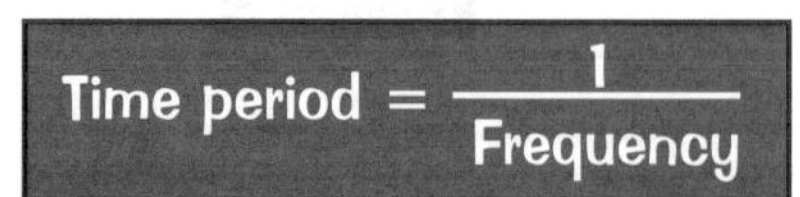

$$\text{Time period} = \frac{1}{\text{Frequency}}$$

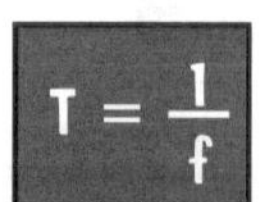

$$T = \frac{1}{f}$$

Where:
T = the period time in seconds (s)
f = frequency of the pendulum in hertz (Hz)

5) The time period of a pendulum depends on its length. The longer the pendulum, the greater the time period. So the shorter the length, the shorter the time period.
6) As well as being using in old-style clocks, pendulums have many other (more fun) uses. For example, playground swings are pendulums. Any fairground rides that swing you back and forth are pendulums too. Hooray for pendulums.

You are feeling very sleepy, verrrrry sleeeeepy...

So there you go, the science behind the dangers of the age-old, time-passing activity of 'stool-swinging'. If the centre of mass of you and the stool falls outside of the stool's base, then you're heading for a fall. Ouch.

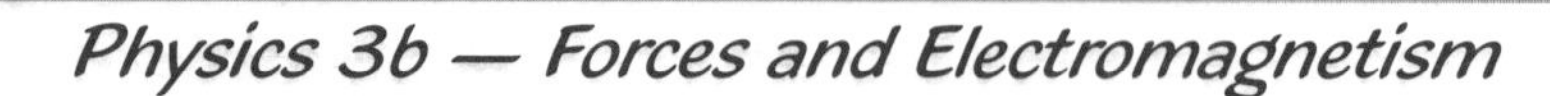

Hydraulics

Oh my word, hydraulics. I have to say that word sounds a little scary, but it's not all that bad really. It's all just about how we can use the properties of liquids to our advantage. Mwahaha.

Liquids are Virtually Incompressible

See page 13 for more on liquids.

1) Liquids are virtually incompressible — you can't squash them, their volume and density stay the same.
2) Because liquids are incompressible and can flow, a force applied to one point in the liquid will be transmitted (passed) to other points in the liquid.
3) Imagine a balloon full of water with a few holes in it. If you squeeze the top of balloon, the water will squirt out of the holes. This shows that force applied to the water at the top of the balloon is transmitted to the water in other parts of the balloon. This also shows that pressure can be transmitted throughout a liquid.

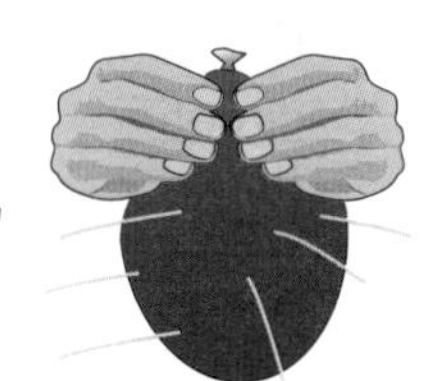

Pressure in a liquid is transmitted equally in all directions.

Pressure and force are linked — see formula below.

Pressure is the Force per Unit Area

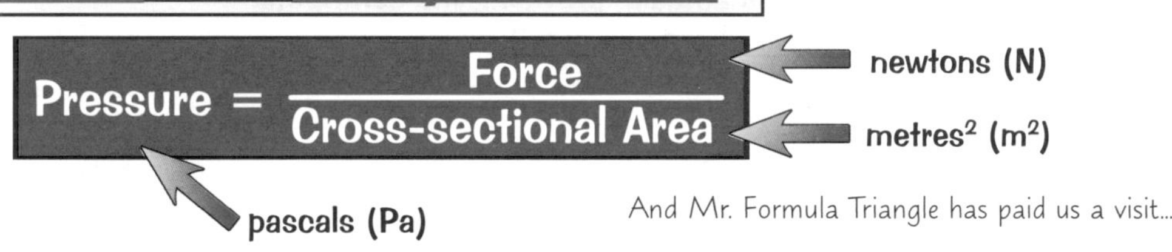

And Mr. Formula Triangle has paid us a visit...

The Pressure in Liquids can be Used in Hydraulic Systems

1) Hydraulic systems are used as force multipliers — they use a small force to produce a bigger force. They do this using liquid and a sneaky trick with cross-sectional areas.
2) The diagram to the right shows a simple hydraulic system.
3) The system has two pistons, one with a smaller cross-sectional area than the other. Pressure is transmitted equally through a liquid — so the pressure at both pistons is the same.
4) Pressure = force ÷ area, so at the 1st piston, a pressure is exerted on the liquid using a small force over a small area. This pressure is transmitted to the 2nd piston.
5) The 2nd piston has a larger area, and so as force = pressure × area, there will be a larger force.
6) Hydraulic systems are used in all sorts of things, e.g. car braking systems, hydraulic car jacks, manufacturing and deployment of landing gear on some aircraft.

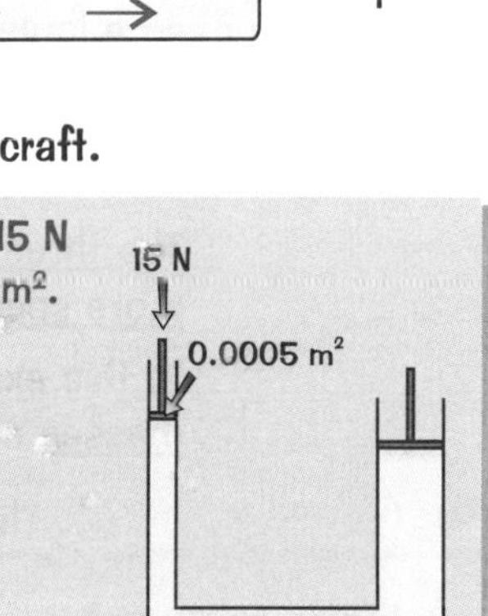

EXAMPLE: To the right is a diagram showing a simple hydraulic system. A force of 15 N is applied to the first piston which has a cross-sectional area of 0.0005 m².
a) Calculate the pressure created on the first piston.
b) Calculate the force acting on the second piston if its cross-sectional area is 0.0012 m².

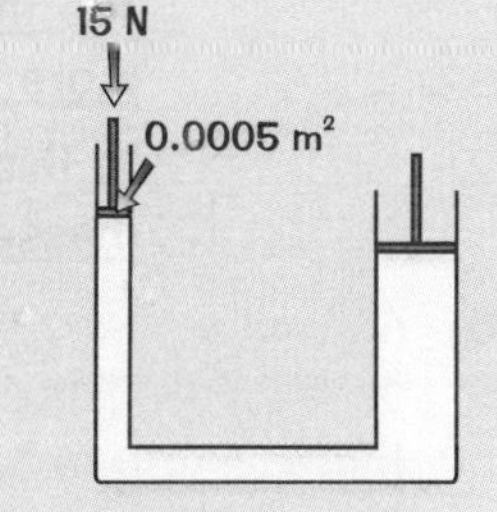

ANSWER:
a) P = F ÷ A = 15 ÷ 0.0005 = 30 000 Pa (or 30 000 N/m²)
b) Pressure at first piston = pressure at second piston, so
F = P × A = 30 000 × 0.0012 = 36 N

With all this talk of hydraulics I'm really feeling under pressure...

See, that wasn't too bad was it? Maybe-ish. Hydraulics is all about using a liquid to make a larger force from a smaller one (or in some cases vice versa). It's all thanks to those hard-to-compress liquids, with their pressure-transmitting capabilities. I'll never look at glass of water in the same way again.

Circular Motion

If it wasn't for circular motion our little planet would just be wandering aimlessly around the universe. And as soon as you launched a satellite, it'd just go flying off into space. Hardly ideal.

Circular Motion — Velocity is Constantly Changing

1) Velocity is both the speed and direction of an object (p.43).
2) If an object is travelling in a circle it is constantly changing direction. This means its velocity is constantly changing (but not its speed) — so the object is accelerating (p.44). This acceleration is towards the centre of the circle.
3) There must be a resultant force acting on the object causing this acceleration (p.48). This force acts towards the centre of the circle.
4) This force that keeps something moving in a circle is called a centripetal force.

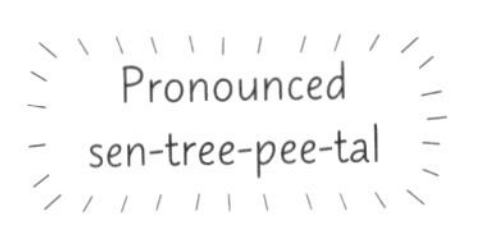

The object's acceleration changes the direction of motion but not the speed.

The force causing the acceleration is always towards the centre of the circle.

In the exam, you could be asked to say which force is actually providing the centripetal force in a given situation. It can be tension, or friction, or even gravity.

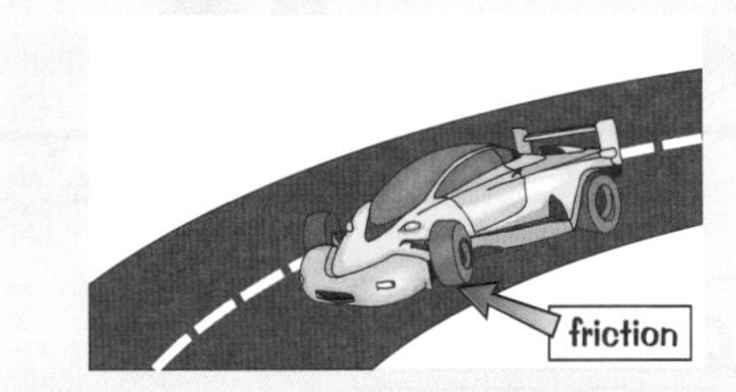

A car going round a bend:
1) Imagine the bend is part of a circle — the centripetal force is towards the centre of the circle.
2) The force is from friction between the car's tyres and the road.

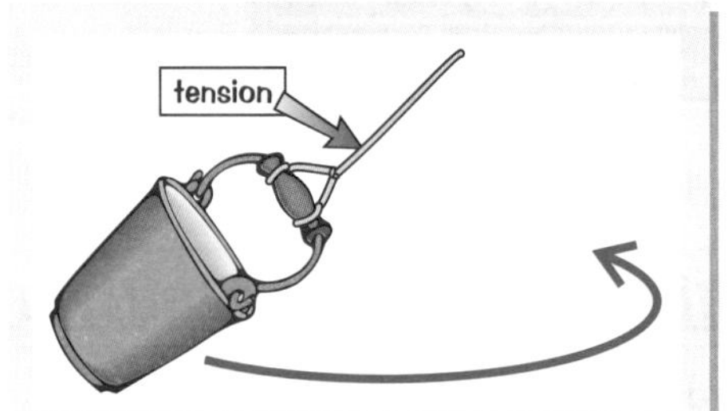

A bucket whirling round on a rope:
The centripetal force comes from tension in the rope. Break the rope, and the bucket flies off at a tangent.

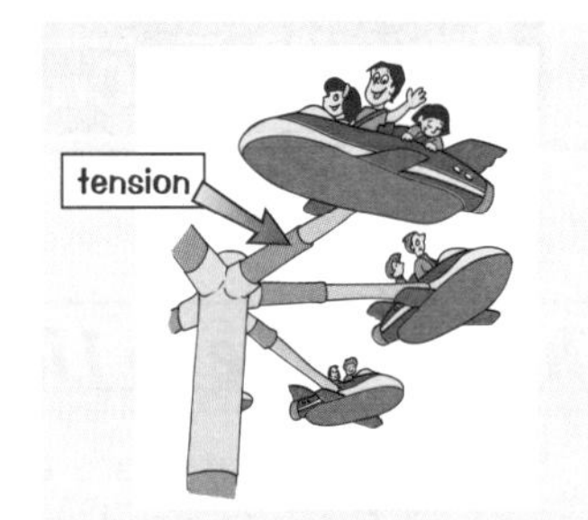

A spinning fairground ride:
The centripetal force comes from tension in the spokes of the ride.

Centripetal Force depends on Mass, Speed and Radius

1) The faster an object's moving, the bigger the centripetal force has to be to keep it moving in a circle.
2) The larger the mass of the object, the bigger the centripetal force has to be to keep it moving in a circle.
3) And you need a larger force to keep something moving in a smaller circle — it has 'more turning' to do.

Example: Two cars are driving at the same speed around the same circular track. One has a mass of 900 kg, the other has a mass of 1200 kg. Which car has the larger centripetal force?

The three things that mean you need a bigger centripetal force are:
more speed, more mass, smaller radius of circle.

In this example, the speed and radius of circle are the same — the only difference is the masses of the cars. So you don't need to calculate anything — you can confidently say:

The 1200 kg car (the heavier one) must have the larger centripetal force.

Circular motion — get round to learning it...

To understand this, you need to learn that constant change in direction means constant acceleration. Velocity is a vector — it has direction, and acceleration is change in velocity. When there's acceleration, there's force (see, easy). Learn what forces can provide centripetal force — e.g. tension and friction.

Magnetic Fields

There's a proper definition of a magnetic field which you really ought to learn:

A MAGNETIC FIELD is a region where MAGNETIC MATERIALS (like iron and steel) and also WIRES CARRYING CURRENTS experience A FORCE acting on them.

Magnetic fields can be represented by field diagrams (e.g. see coil of wire diagram below).
The arrows on the field lines always point FROM THE NORTH POLE of the magnet TO THE SOUTH POLE!

The Magnetic Field Round a Current-Carrying Wire

1) When a current flows through a wire, a magnetic field is created around the wire.
2) The field is made up of concentric circles with the wire in the centre.

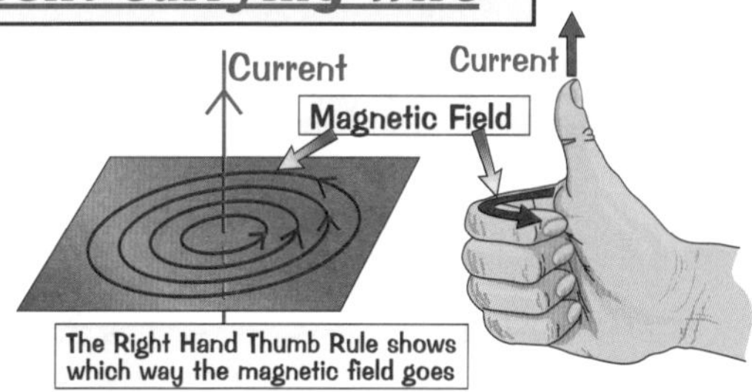

The Magnetic Field Round a Coil of Wire

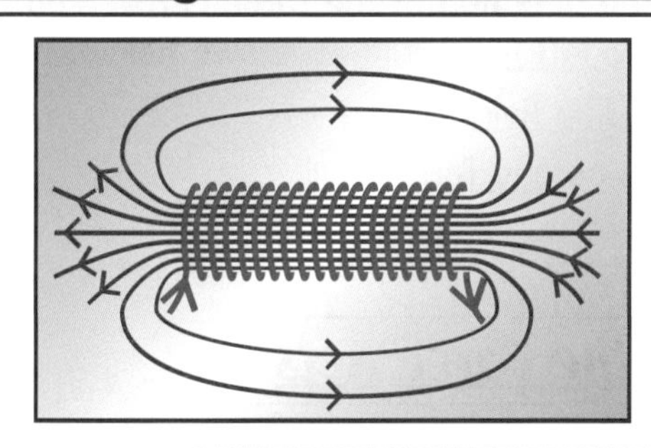

1) The magnetic field inside a coil of wire (a solenoid) is strong and uniform.
2) Outside the coil, the magnetic field is just like the one round a bar magnet.
3) You can increase the strength of the magnetic field around a solenoid by adding a magnetically "soft" iron core through the middle of the coil. It's then called an ELECTROMAGNET.

A magnetically soft material magnetises and demagnetises very easily. So, as soon as you turn off the current through the solenoid, the magnetic field disappears — the iron doesn't stay magnetised. This is what makes it useful for something that needs to be able to switch its magnetism on and off (see below).

Electromagnets are Useful as Their Magnetism can be Switched Off

An electromagnet must be constantly supplied with current — as that's what produces the magnetic field. So if the current stops, then it stops being magnetic. Magnets you can switch off at your whim can be really useful...

Example: Cranes Used for Lifting Iron and Steel

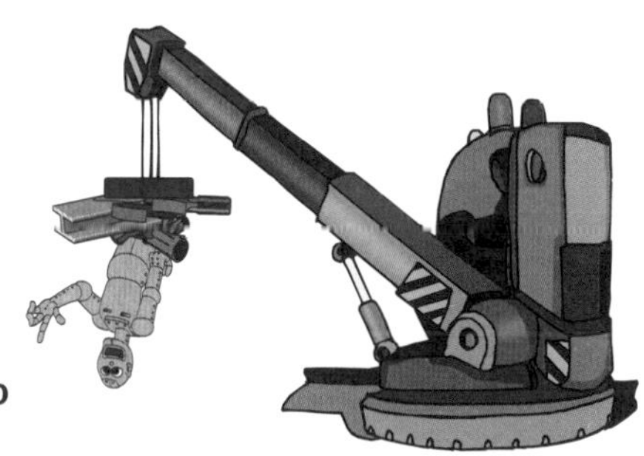

1) Magnets can be used to attract and pick up things made from magnetic materials like iron and steel.
2) Electromagnets are used in some cranes, e.g. in scrap yards and steel works.
3) If an ordinary magnet was used, the crane would be able to pick up the cars etc., but then wouldn't let it go. Which isn't very helpful.
4) Using an electromagnetic means the magnet can be switched on when you want to and attract and pick stuff up, then switched off when you want to drop it. Which is far more useful.

I'm magnetically soft — I always cry when electromagnets are turned off...

Electromagnets pop up in lots of different places. For example, they're used in electric bells, car ignition circuits and some security doors. Electromagnets aren't all the same strength though, that wouldn't work. How strong they are depends on stuff like the number of turns of wire there are and the size of current going through the wire.

The Motor Effect

Passing an electric current through a wire produces a magnetic field around the wire (p.97). If you put that wire into a magnetic field, you have two magnetic fields combining, which puts a force on the wire (generally).

A Current in a Magnetic Field Experiences a Force

The force experienced by a current-carrying wire in a magnetic field is known as the motor effect.

1) The two tests below demonstrate the force on a current-carrying wire placed in a magnetic field. The force gets bigger if either the current or the magnetic field is made bigger.

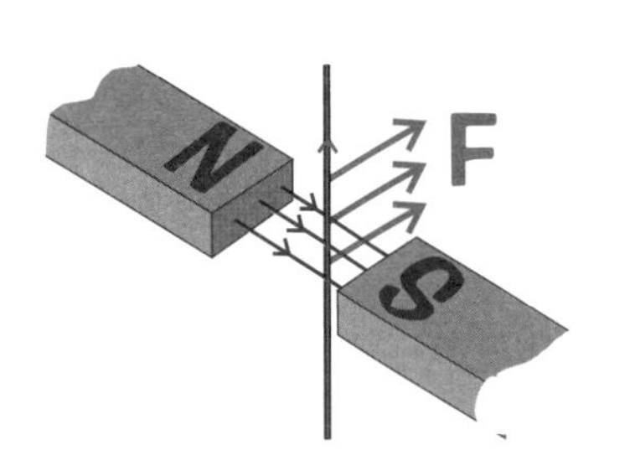

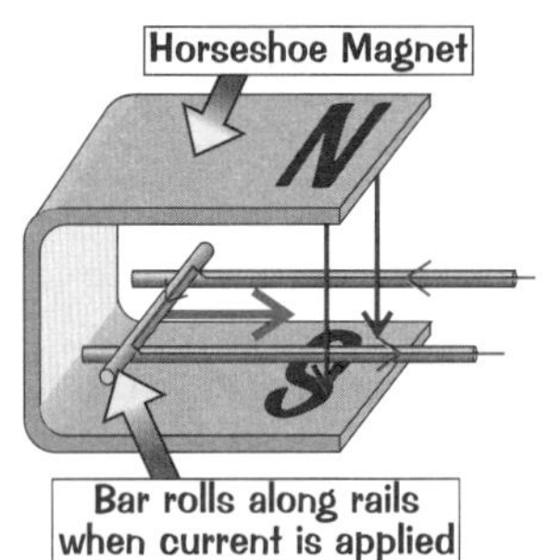

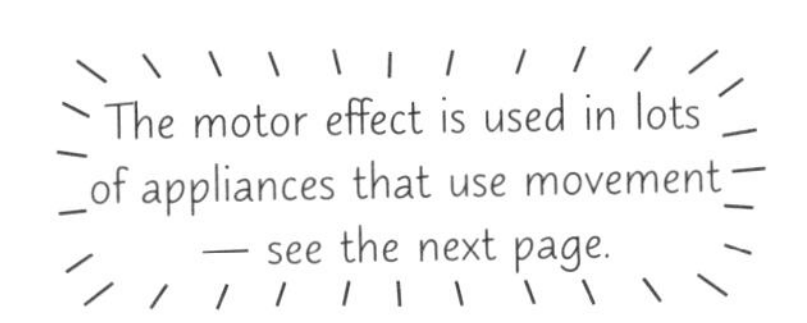

2) Note that in both cases the force on the wire is at 90° to both the wire and to the magnetic field.
3) If the direction of the current or magnetic field is reversed, then the direction of the force is reversed too. You can always predict which way the force will act using Fleming's left hand rule as shown below.
4) To experience the full force, the wire has to be at 90° to the magnetic field.
5) If the wire runs along parallel to the magnetic field it won't experience any force at all.
6) At angles in between it'll feel some force.

Fleming's Left Hand Rule Tells You Which Way the Force Acts

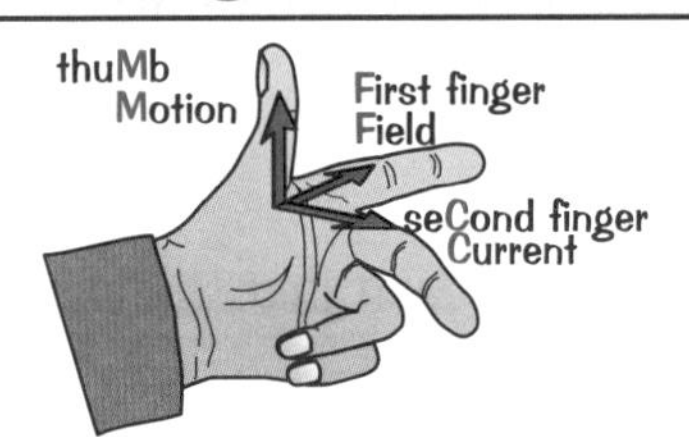

1) They could test if you can do this, so practise it.
2) Using your left hand, point your First finger in the direction of the Field and your seCond finger in the direction of the Current.
3) Your thuMb will then point in the direction of the force (Motion).

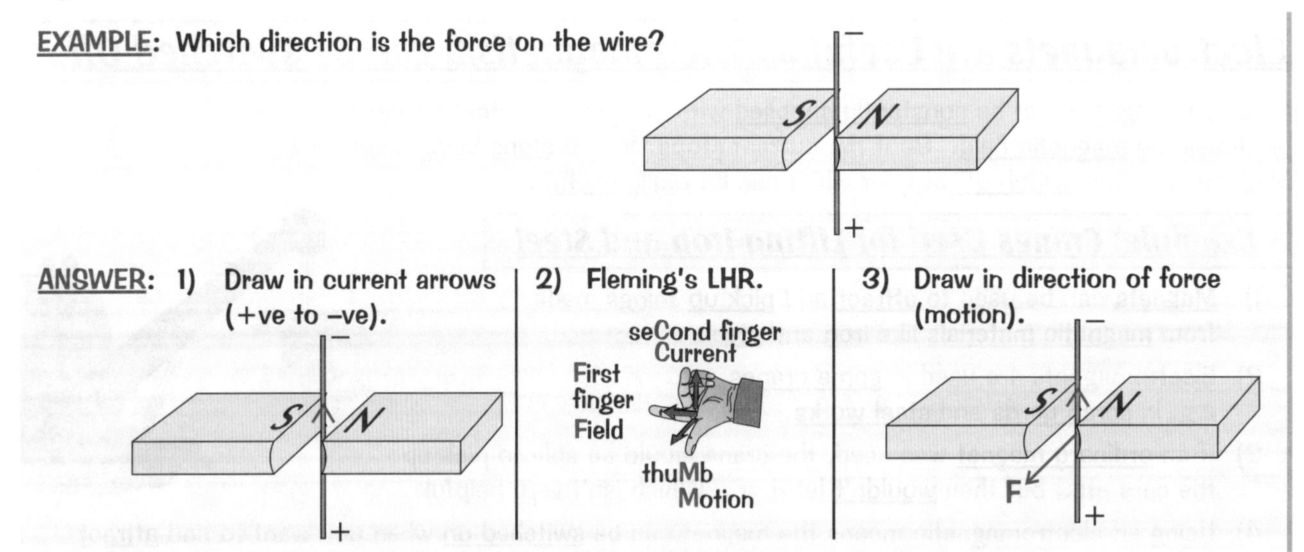

Remember the Left Hand Rule for Motors — drive on the left...

Always remember that it's the LEFT hand rule. If you whip your right hand out in the exam and start looking at the fingers, you'll get it WRONG. Remember that magnetic fields go from north to south, not south to north. And yes, it seems weird that magnets move wires, but that's physics for you.

The Simple Electric Motor

Electric motors use the motor effect (p.98) to get them (and keep them) moving. This is one of the favourite exam topics of all time. Read it. Understand it. Learn it. Lecture over.

The Simple Electric Motor

2 Factors which Speed it up:

1) More CURRENT
2) STRONGER MAGNETIC FIELD

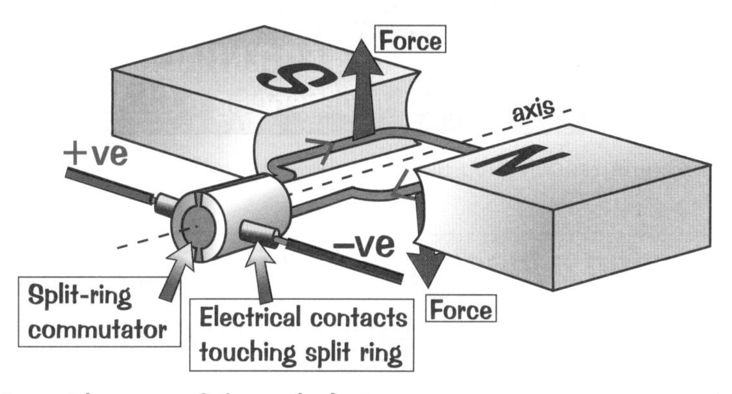

1) The diagram shows the forces acting on the two side arms of the coil of wire.
2) These forces are just the usual forces which act on any current in a magnetic field.
3) Because the coil is on a spindle and the forces act one up and one down, it rotates.
4) The split-ring commutator is a clever way of "swapping the contacts every half turn to keep the motor rotating in the same direction". (Learn that statement because they might ask you.)
5) The direction of the motor can be reversed either by swapping the polarity of the direct current (DC) supply or swapping the magnetic poles over.

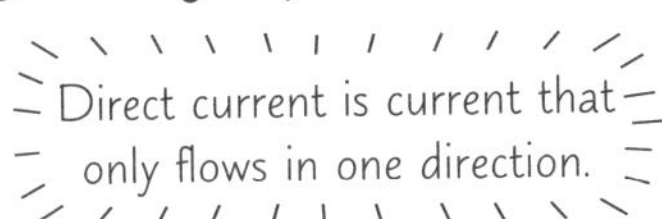

EXAMPLE: Is the coil turning clockwise or anticlockwise?

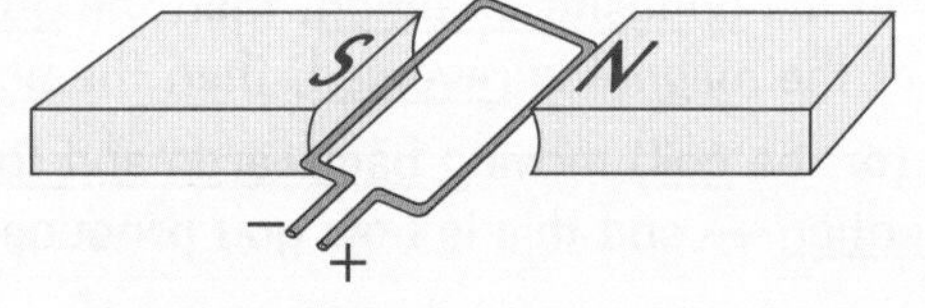

ANSWER: 1) Draw in current arrows (+ve to –ve).

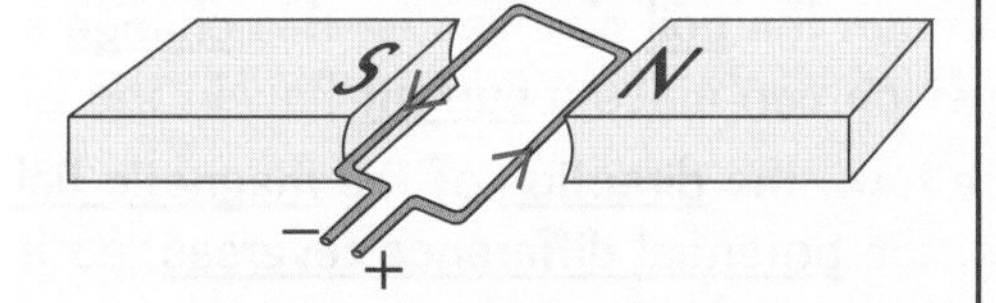

2) Fleming's LHR on one side arm (I've used the right hand arm).

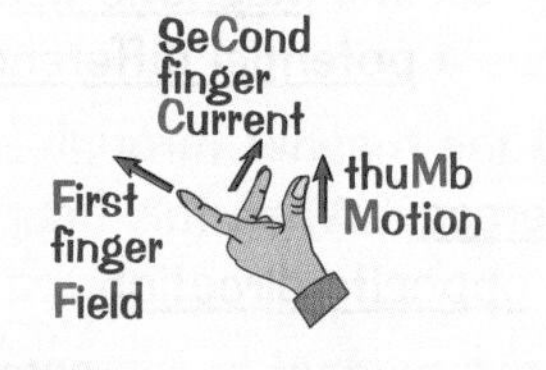

3) Draw in direction of force (motion).

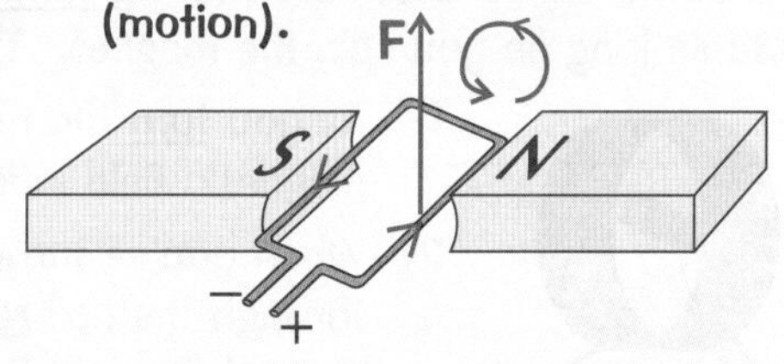

So — the coil is turning anticlockwise.

Electric Motors are used in: CD Players, Food Mixers, Fan Heaters... ...Fans, Printers, Drills, Hair Dryers, Cement Mixers, etc.

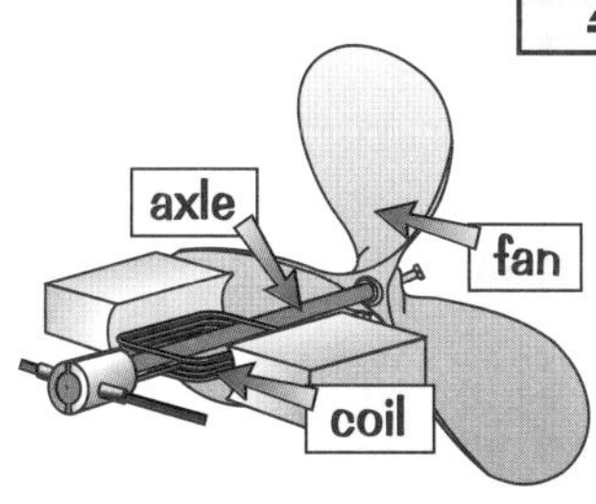

1) Link the coil to an axle, and the axle spins round.
2) In the diagram there's a fan attached to the axle, but you can stick almost anything on a motor axle and make it spin round.
3) For example, in a food mixer, the axle's attached to a blade or whisks. In a CD player the axle's attached to the bit you sit the CD on. Fan heaters and hair dryers have an electric heater as well as a fan.

Hey, don't call my electric motor simple...

Those electric motors get everywhere. Life'd be a much sadder place without them and my hair would look even more ridiculous. It's all thanks to the motor effect. Make sure you know some appliances that use the motor effect and can describe how it helps to make each appliance work. It's the sorta thing that might be on the exam.

Electromagnetic Induction

Sounds terrifying. Well, sure it's quite mysterious, but it isn't that complicated:

ELECTROMAGNETIC INDUCTION:
The creation of a POTENTIAL DIFFERENCE across a conductor which is experiencing a CHANGE IN MAGNETIC FIELD.

Remember — potential difference is just another name for voltage.

For some reason they use the word "induction" rather than "creation", but it amounts to the same thing.

Moving a Magnet in a Coil of Wire Induces a Voltage

If the conductor is part of a complete circuit, a current will flow.

1) Electromagnetic induction means creating a potential difference across the ends of a conductor (e.g. a wire).
2) You can do this by moving a magnet in a coil of wire or moving an electrical conductor in a magnetic field ("cutting" magnetic field lines). Shifting the magnet from side to side creates a little "blip" of current.

A few examples of electromagnetic induction:

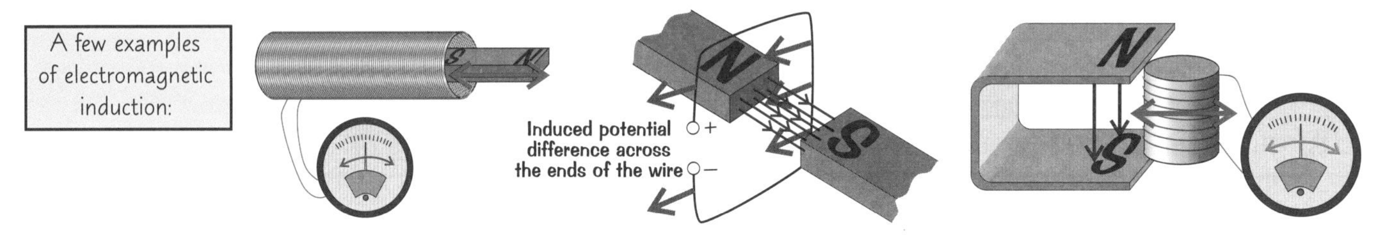

3) If you move the magnet in the opposite direction, then the potential difference/current will be reversed too. Likewise, if the polarity of the magnet is reversed, then the potential difference/current will be reversed too.
4) If you keep the magnet (or the coil) moving backwards and forwards, you produce a potential difference that keeps swapping direction — and this is how you produce an alternating current (AC) — see p.66.

You can create the same effect by turning a magnet end to end in a coil, to create a current that lasts as long as you spin the magnet. This is how generators work (see below for an example).

1) As you turn the magnet, the magnetic field through the coil changes — this change in the magnetic field induces a potential difference, which can make a current flow in the wire.
2) When you've turned the magnet through half a turn, the direction of the magnetic field through the coil reverses. When this happens, the potential difference reverses, so the current flows in the opposite direction around the coil of wire.
3) If you keep turning the magnet in the same direction — always clockwise, say — then the potential difference will keep on reversing every half turn and you'll get an AC current.

Some Appliances use Electromagnetic Induction to Generate a Current

Example: Dynamos

Dynamos are often used on bikes to power the lights. The cog wheel at the top is positioned so that it touches one of the wheels. As the wheel moves round, it turns the cog which is attached to the magnet. This creates an AC current to power the lights.

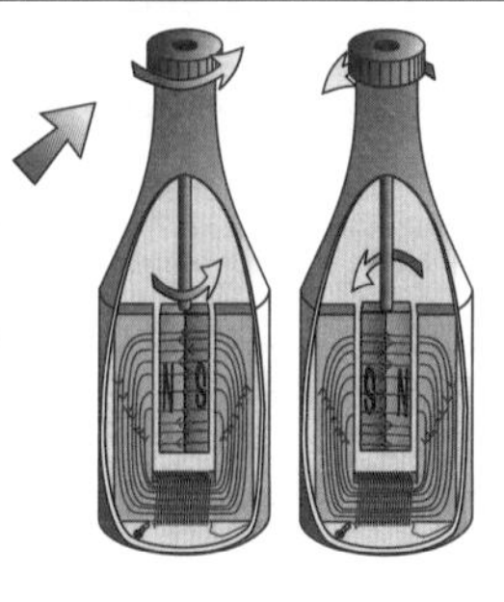

EM induction — works whether the coil or the field is moving...

"Electromagnetic Induction" gets my vote for "Definitely Most Tricky Topic". If it wasn't so important maybe you wouldn't have to bother learning it. The trouble is, this is how most of our electricity is generated, whether it's in a coal-fired power station or a wind turbine. Well done electromagnetic induction. Well done.

Transformers

Transformers use electromagnetic induction to change potential difference (p.d.). So they will only work on AC.

Transformers Change the p.d. — but only AC p.d.

There are a few different types of transformer. The two you need to know about are step-up transformers and step-down transformers. They both have two coils, the primary and the secondary, joined with an iron core.

STEP-UP TRANSFORMERS step the voltage up. They have more turns on the secondary coil than the primary coil.

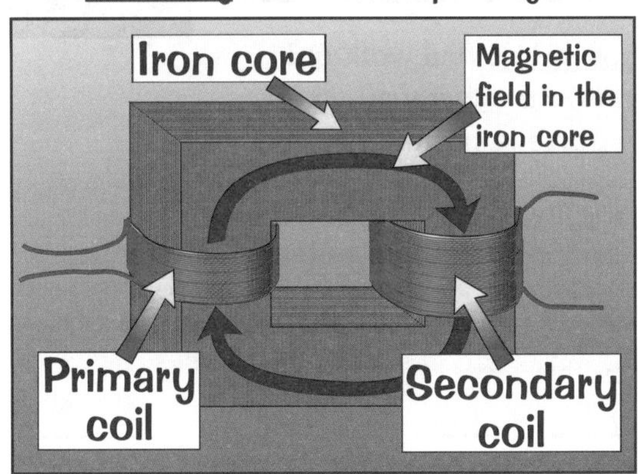

STEP-DOWN TRANSFORMERS step the voltage down. They have more turns on the primary coil than the secondary.

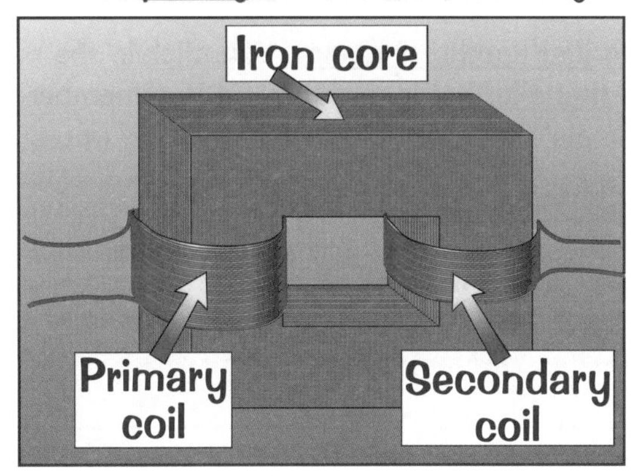

Transformers Work by Electromagnetic Induction

1) The primary coil produces a magnetic field which stays within the iron core. This means nearly all of it passes through the secondary coil and hardly any is lost.
2) Because there is alternating current (AC) in the primary coil, the field in the iron core is constantly changing direction (100 times a second if it's at 50 Hz) — i.e. it is a changing magnetic field.
3) This rapidly changing magnetic field is then felt by the secondary coil.
4) The changing field induces an alternating potential difference across the secondary coil (with the same frequency as the alternating current in the primary) — electromagnetic induction of a potential difference in fact.
5) The relative number of turns on the two coils determines whether the potential difference induced in the secondary coil is greater or less than the potential difference in the primary.
6) In a step-up transformer, the p.d. across the secondary coil is greater than the p.d. across the primary coil.
7) In a step-down transformer, the p.d. across the secondary coil is less than the p.d. across the primary coil.
8) If you supplied DC to the primary, you'd get nothing out of the secondary at all. Sure, there'd still be a magnetic field in the iron core, but it wouldn't be constantly changing, so there'd be no induction in the secondary because you need a changing field to induce a potential difference. Don't you! So don't forget it — transformers only work with AC. They won't work with DC at all.

The Iron Core Carries Magnetic Field, Not Current

1) The iron core is purely for transferring the changing magnetic field from the primary coil to the secondary.
2) No electricity flows round the iron core.

The ubiquitous Iron Core — where would we be without it...

Transformers only work with AC. I'll say that again. Transformers only work with AC. Prevent disaster in the exam by remembering the fact that transformers only work with AC. Now that's out of the way, I recommend you learn the details and the diagrams, then cover the page and scribble them down.

Transformers

Ah, more about transformers. And as per usual, some equations to learn too. I don't like change.

The Transformer Equation — use it Either Way Up

You can calculate the output potential difference from a transformer if you know the input potential difference and the number of turns on each coil.

$$\frac{\text{Potential Difference across Primary Coil}}{\text{Potential Difference across Secondary Coil}} = \frac{\text{Number of turns on Primary Coil}}{\text{Number of turns on Secondary Coil}}$$

$$\frac{V_P}{V_S} = \frac{n_P}{n_S}$$

or

$$\frac{V_S}{V_P} = \frac{n_S}{n_P}$$

Well, it's just another formula. You stick in the numbers you've got and work out the one that's left. It's really useful to remember you can write it either way up — this example's much trickier algebra-wise if you start with V_S on the bottom...

EXAMPLE: A transformer has 40 turns on the primary and 800 on the secondary. If the input potential difference is 1000 V, find the output potential difference.

ANSWER: $V_S/V_P = n_S/n_P$ so $V_S/1000 = 800/40$ $V_S = 1000 \times (800/40) = $ 20 000 V

Transformers are Nearly 100% Efficient So "Power In = Power Out"

The formula for power supplied is: Power = Current × Potential Difference or: $P = I \times V$ (see page 70).

So you can write electrical power input = electrical power output as:

$$V_pI_p = V_sI_s$$

V_p = p.d. across primary coil (V)
I_p = current in the primary coil (A)
V_s = p.d. across secondary coil (V)
I_s = current in the secondary coil (A)

EXAMPLE: A transformer in a travel adaptor steps up a 110 V AC mains electricity supply to the 230 V needed for a hair dryer. The current through the hair dryer is 5 A. If the transformer is 100% efficient, calculate how much current is drawn by the transformer from the mains supply.

ANSWER: $V_p \times I_p = V_s \times I_s$ so $110 \times I_p = 230 \times 5$ $I_p = (230 \times 5) \div 110 = $ 10.5 A

Switch Mode Transformers are used in Chargers and Power Supplies

1) Switch mode transformers are a type of transformer that operate at higher frequencies than traditional transformers.
2) They usually operate at between 50 kHz and 200 kHz.
3) Because they work at higher frequencies, they can be made much lighter and smaller than traditional transformers that work from a 50 Hz mains supply.
4) This makes them more useful in things like mobile phone chargers and power supplies, e.g. for laptops.
5) Switch mode transformers are more efficient than other types of transformer. They use very little power when they're switched on but no load (the thing you're charging or powering) is applied, e.g. if you've left your phone charger plugged in but haven't attached your phone.

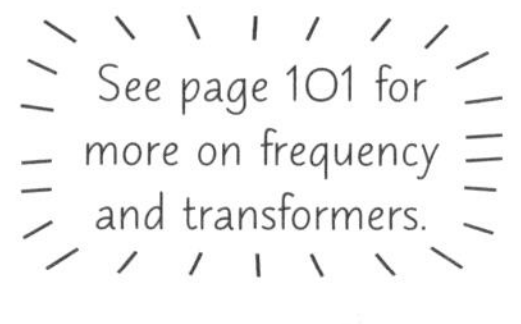
See page 101 for more on frequency and transformers.

Which transformer do you need to enslave the universe — Megatron...

You'll need to practise with those tricky equations. They're unusual because they can't be put into formula triangles, but other than that the method is the same — stick in the numbers. Just practise.

Revision Summary for Physics 3b

There's only one way to check you know it all. Sorry.

1) Sarah is levering the lid off a can of paint using a screwdriver. She places the tip of the 20 cm long screwdriver under the can's lid and applies a force of 10 N on the end of the screwdriver's handle. Suggest two ways that Sarah could increase the moment about the pivot point (the side of the can).
2) Describe two different ways of finding the centre of mass of a rectangular playing card.
3)* Arthur weighs 600 N and is sitting on a seesaw 1.5 m from the pivot point. His friend Caroline weighs 450 N and sits on the seesaw so that it balances. How far from the pivot point is Caroline sitting?
4) Give three situations where you use a simple lever.
5) Give two features of a Bunsen burner that make it difficult to tip over.
6)* Calculate the time period of a pendulum swinging with a frequency of 10 Hz.
7)* A force of 20 N is applied to a piston in a hydraulic system with a cross-sectional area 0.25 m^2. Calculate the pressure applied to the piston.
8) Give one use of a hydraulic system.
9) A cyclist is moving at a constant speed of 5 m/s around a circular track.
   a) Is the cyclist accelerating? Explain your answer.
   b) What force keeps the cyclist travelling in a circle? Where does this force come from?
   c) What will happen to the size of this force if the same cyclist travels at a constant speed of 5 m/s around a different circular track that has a larger radius?
10) What is an electromagnet? Describe one use of electromagnets. Explain why they're good for this job.
11) Describe what happens to a current-carrying wire when it is placed in a magnetic field.
12) Describe the three details of Fleming's left hand rule. What is it used for?
13) The diagrams show a simple electric motor. The coil is turning clockwise. Which diagram, A or B, shows the correct polarity of the magnets?

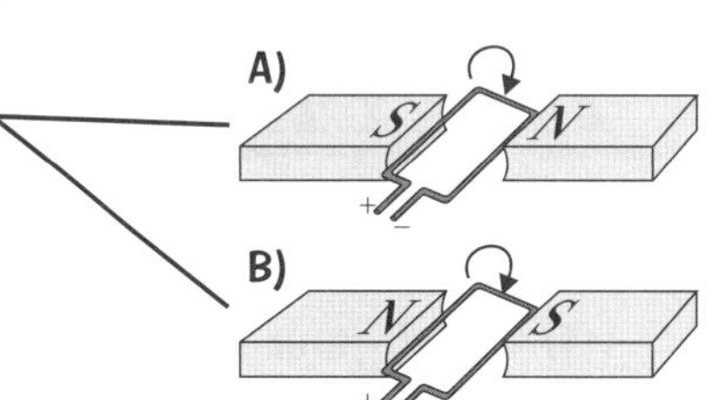

14) Give the definition of electromagnetic induction.
15) Describe two ways in which you could induce a potential difference using a wire and a magnet.
16) Describe how a dynamo works.
17) Sketch two types of transformer and explain the differences between them.
18) An engineering executive is travelling from the USA to Italy and taking a computer monitor with him. In the USA, domestic electricity is 110 V AC, and in Italy it's 230 V AC. What kind of transformer would the engineering executive need to plug his monitor into?
19) Explain how a transformer works and why transformers only work on AC voltage.
20) Write down the transformer equation.
21)* A transformer has 20 turns on the primary coil and 600 on the secondary coil. If the input potential difference is 9 V, find the output potential difference.
22)* A transformer steps down 230 V from the mains supply to the 130 V needed for an appliance. If the transformer draws 2 A from the mains supply, calculate how much current goes through the appliance.
23) Give two advantages of switch mode transformers over traditional transformers.

* Answers on page 108.

The Perfect Cup of Tea

The making and drinking of tea are important life skills. It's not something that will crop up in the exam, but it is something that will make your revision much easier. So here's a guide to making the perfect cuppa...

1) Choose the Right Mug

A good mug is an essential part of the tea drinking experience, but choosing the right vessel for your tea can be tricky. Here's a guide to choosing your mug:

Some bad mugs:

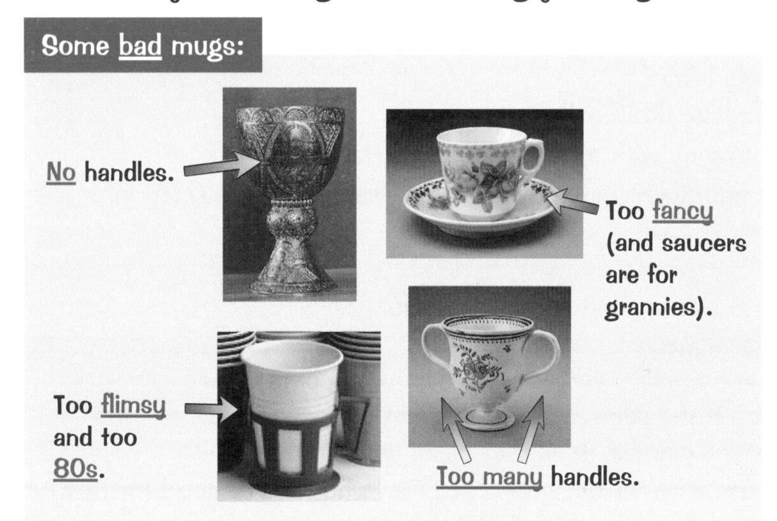

The perfect mug:

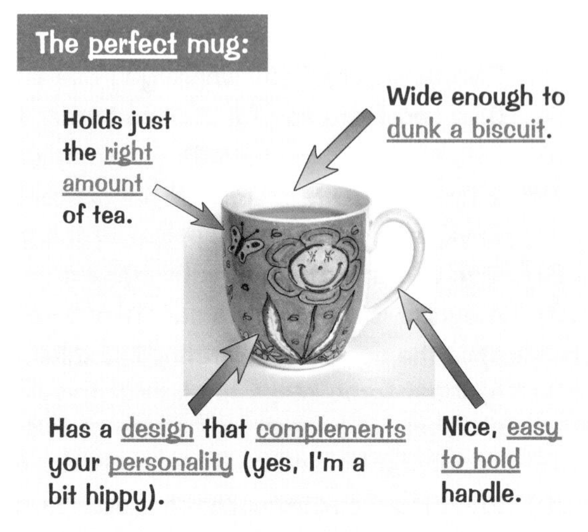

2) Get Some Water and Boil It

For a really great brew follow these easy step-by-step instructions:

1) First, pour some water into a kettle and switch it on. (Check it's switched on at the wall too.)
2) Let the kettle boil. While you're waiting, see what's on TV later and check your belly button for fluff. Oh, and put a tea bag in a mug.
3) Once the kettle has boiled, pour the water into the mug.
4) Mash the tea bag about a bit with a spoon. Remove the tea bag.
5) Add a splash of milk (and a lump of sugar or two if you're feeling naughty).

Top tea tip no. 23: why not ask your mum if she wants a cup too?

Note: some people may tell you to add the milk before the tea. Scientists have recently confirmed that this is nonsense.

3) Sit Back and Relax

Now this is important — once you've made your cuppa:

1) Have a quick rummage in the kitchen cupboards for a cheeky biscuit. (Custard creams are best — steer clear of any ginger biscuits — they're evil.)
2) Find your favourite armchair/beanbag. Move the cat.
3) Sit back and enjoy your mug of tea. You've earned it.

Phew — time for a brew I reckon...

It's best to ignore what other people say about making cups of tea and follow this method. Trust me, this is the most definitive and effective method. If you don't do it this way, you'll have a shoddy drinking experience. There, you've been warned. Now go and get the kettle on. Mine's milk and two sugars...

Index

Index

Index

Answers

Efficiency of Machines Top Tip (page 20)

TV: 0.0318 or 3.18%

Loudspeaker: 0.0143 or 1.43%

Revision Summary for Physics 1a (page 25)

15) 40 years

20) 90 kJ

26) 70% or 0.7

27) a) 80 J b) 20 J c) 0.8 or 80%

29) 0.125 kWh

Wave Basics Top Tip (page 34)

2735 m/s

Revision Summary for Physics 1b (page 42)

15) 150 m/s

Revision Summary for Physics 2a (page 57)

5) $a = (v - u) \div t$
$a = (14 - 0) \div 0.4 = 35$ m/s^2

11) $F = ma$
$a = F \div m = 30 \div 4 = 7.5$ m/s^2

12) Downward force of gravity on skydiver:
$F = m \times a = 75 \times 10 = 750$ N.
Resultant force at 80 mph:
$F = 750 - 650 = 100$ N downwards.
Resultant acceleration:
$a = F \div m = 100 \div 75 = 1.33$ m/s^2

13) 120 N

16) Work done = force × distance.
$W = 535 \times 12 = 6420$ J

17) a) $Ep = m \times g \times h = 4 \times 10 \times 30 = 1200$ J
b) $Ep = 1200 \div 2 = 600$ J

18) $Ek = \frac{1}{2} \times m \times v^2$
$Ek = \frac{1}{2} \times 78 \times 23^2 = 20631$ J

19) Ek just as she hits the ground = Ep at the top.
($g = 10$ N/kg)
So $Ek = m \times g \times h = 78 \times 10 \times 20 = 15\,600$ J

20) Ek transferred = work done by brakes
$\frac{1}{2} \times m \times v^2 = F \times d$
$\frac{1}{2} \times 1000 \times 2^2 = 395 \times d$
$d = 2000 \div 395 = 5.1$ m
The car would come to stop in 5.1 m, so no, he can't avoid hitting the sheep.

23) $P = (m \times g \times h) \div t$
($g = 10$ N/kg)
$P = (78 \times 10 \times 20) \div 16.5 = 945$ W

Revision Summary for Physics 2b (page 80)

3) $I = Q \div t$, so $I = 240 \div (1 \times 60) = 4$ A

8) $V = I \times R$, so $V = 2 \times 0.6 = 1.2$ V

11) a) Current is the same everywhere in the circuit and resistance adds up in a series circuit.
Total resistance $= 4 + 6 = 10\ \Omega$
$I = V \div R = 12 \div 10 = 1.2$ A
b) Potential difference is shared between the bulbs. $V = I \times R = 1.2 \times 6 = 7.2$ V
c) In parallel, the potential difference is the same over each branch of the circuit and is equal to the supply potential difference, therefore the potential difference over either bulb = 12 V.

12) $f = 1 \div T$, so $f = 1 \div 0.08 = 12.5$ Hz

17) $P = E \div t$, $E = P \times t$
Hair straighteners: $E = 45 \times (5 \times 60) = 13\,500$ J
Hair dryer: $E = 105 \times (2 \times 60) = 12\,600$ J
The hair straighteners use more energy.

18) $P = I \times V$, $I = P \div V$
a) $I = 1100 \div 230 = 4.8$ A, so use a 5 A fuse.
b) $I = 2000 \div 230 = 8.7$ A, so use a 13 A fuse.

19) $E = Q \times V$
$E = 530 \times 6 = 3180$ J

Revision Summary for Physics 3a (page 91)

7) $s = v \times t$
$s = 1000 \times 0.00004 = 0.04$ m
so thickness of fat $= 0.04 \div 2 = 0.02$ m $= 2$ cm

12) Refractive index (n) = sin i ÷ sin r
$i = 27°$, $r = 18°$
$n = \sin 27 \div \sin 18 = 1.47$ (to 2 d.p.)

16) Magnification = image height ÷ object height
image height = 4.5, object height = 1.5
Magnification $= 4.5 \div 1.5 = 3$

17) Power = 1 ÷ focal length
focal length = 10 cm = 0.1 m
Power $= 1 \div 0.1 = 10$ D

Revision Summary for Physics 3b (page 103)

3) $1.5 \times 600 = d \times 450$, so
$d = 900 \div 450 = 2$ m

6) $T = 1 \div f$
$T = 1 \div 10 = 0.1$ s

7) $P = F \div A$
$P = 20 \div 0.25 = 80$ Pa

21) $V_s \div V_p = n_s \div n_p$ so $V_s = (600 \div 20) \times 9$
$= 30 \times 9 = 270$ V

22) $V_p \times I_p = V_s \times I_s$
so $I_s = (230 \times 2) \div 130 = 3.5$ A